Casting Processes and Modelling of Metallic Materials

Edited by Zakaria Abdallah and Nada Aldoumani

Published in London, United Kingdom

IntechOpen

Supporting open minds since 2005

Casting Processes and Modelling of Metallic Materials
http://dx.doi.org/10.5772/intechopen.91879
Edited by Zakaria Abdallah and Nada Aldoumani

Contributors
Rodolfo Morales, Maria Guadalupe González-Solórzano, Jafeth Rodríguez Ávila, Carlos-Rodrigo Muñiz-Valdez, Alfonso Nájera Bastida, John Campbell, Fawzy Samuel, Agnes Marie Samuel, Herbert Doty, Mohamed Abdelaziz, Dominque Gagnon, Nawaz Mahomed, Kenichi Yano, Yuto Takagi, Masahiro Inagaki, Goutam Sutradhar, Partha Haldar, Anna Gowsalya Lucas, Mahboob E Afshan, Hrushikesh Sarangi, Upendra Kumar Mohanty

First published in London, United Kingdom, 2021 by IntechOpen
IntechOpen is the global imprint of INTECHOPEN LIMITED, registered in England and Wales, registration number: 11086078, 5 Princes Gate Court, London, SW7 2QJ, United Kingdom
Printed in Croatia

British Library Cataloguing-in-Publication Data
A catalogue record for this book is available from the British Library

Additional hard and PDF copies can be obtained from orders@intechopen.com

Casting Processes and Modelling of Metallic Materials
Edited by Zakaria Abdallah and Nada Aldoumani
p. cm.
Print ISBN 978-1-83968-431-9
Online ISBN 978-1-83968-432-6
eBook (PDF) ISBN 978-1-83968-433-3

For EU product safety concerns:
IN TECH d.o.o., Prolaz Marije Krucifikse Kozulić 3, 51000 Rijeka, Croatia, info@intechopen.com or visit our website at intechopen.com.

Meet the editors

Dr. Zakaria Abdallah is the Principal and Lead Researcher of fracture and fatigue in the Steel and Metals Institute (SaMI) at Swansea University. This research institute acts as the research and development center for the various steel and metal industries in the UK. The role of Dr. Abdallah involves the in-house manufacturing of steel as well as testing of the mechanical properties and fracture characteristics under various testing conditions. Dr. Abdallah has also worked at the Rolls-Royce University Technology Centre at Swansea University. This role involved research and development of the high temperature alloys utilized in aeroengines. This work provided extensive expertise in nickel, titanium, and steel alloys employed in gas turbines. The power gearbox in the Rolls-Royce Ultrafan® aeroengine was one of the major projects on which Dr. Abdallah has been working for several years. In the meantime, Dr. Abdallah has worked as a consultant for various industries, e.g. Airbus, TIMET, ETD, Rolls-Royce, in the UK under the umbrella of Swansea Materials Research and Testing (SMaRT) Ltd. Dr. Abdallah leads, or has led, several modules at Swansea University, e.g. advanced materials, fundamentals of materials, math for materials engineers, design against fatigue, etc. Dr. Abdallah is the supervisor of several undergraduate and postgraduate (MSc and Ph.D.) students working on various projects related to steel and metallic alloys. Dr. Abdallah has a very good track record of publications in internationally-recognized and peer-reviewed journals as well as international conferences and open access books. His research interests include: steel and metals, composite materials, materials characterization, creep and fatigue, life predictions of materials, thermo-mechanical testing, and heat treatment.

Dr. Nada Aldoumani is the Director of Innovative Technologies & Engineering Consultancy House (ITECH) Ltd., a registered business in the UK with headquarters based in London. Dr. Aldoumani provides services in metallic structures that make a major breakthrough in the field of uncertainty quantification of materials. In her business, Dr. Aldoumani is enhancing her consultancy expertise by providing services to customers worldwide. This includes a finite element (FE) modeling and engineering solutions for newly designed products with the aid of uncertainty quantification and problem solving. Dr. Aldoumani was a fellow researcher at Swansea University with a structural engineering background. Dr. Aldoumani is an expert in composite materials for structural applications alongside extensive expertise in finite element (FE) modeling, ANSYS, PDL, ACP, and MATLAB coding. Dr. Aldoumani is also a pioneer in the era of uncertainty quantification in composite structures and adhesively bonded materials, in particular, those employed in the aerospace industry. During her career, Dr. Aldoumani has developed a novel approach that automates ANSYS through a MATLAB code that is capable of running thousands of trials to obtain probabilistic Gaussian distributions. Her work is now published in internationally recognized and peer-reviewed journals.

Contents

Preface

It is a great privilege to be working in the field of materials science and engineering, which I feel is one of the most expanding fields in engineering. Materials are used in all applications around us and this includes (but is not limited to) buildings, offices, cars, aircrafts, ships, rockets, etc. In other words, this field is important to all humans and that is why I personally consider it one of the most interesting fields. I have worked over the years in all aspects of materials engineering starting from the development, alloying, casting, mechanical testing and characterization fields, which have given me a wide expertise in materials engineering. I always wanted to put my efforts into action by gathering research from all over the globe to contribute to the field and support other researchers who thrive to develop materials for the various applications. As a fruitful effort, this book is presented to the research community, in particular those working on the casting of metallic alloys, as a guide and reference. The book discusses the aspects of physical casting, modelling, concerns, and mitigation plans to resolve common issues. I am very grateful and thankful to all authors for their valuable efforts in making this project successful and publishable. The work with IntechOpen is outstanding and I will work with them on other projects once I am given the chance in future. Thank you for reading this book and I hope you have enjoyed its content.

Dr. Zak Abdallah
Steel and Metals Institute (SaMI),
Swansea University,
Swansea, Wales, United Kingdom

Dr. Nada Aldoumani
Swansea University,
Swansea, Wales, United Kingdom

Chapter 1

Perspective Chapter: A Personal Overview of Casting Processes

John Campbell

Abstract

Casting processes are reviewed from the point of view of the type of defects they produce and their consequential properties of the castings they produce, particularly resistance to fracture, and therefore, their reliability in service. The ingot casting of steels is criticized for unnecessary degradation of the steel. The fundamental problems of continuous casting of aluminum alloys and steels are seen to be lying in inattention to the details of the processes. Vacuum casting, particularly vacuum arc remelting, as currently executed, is seen to be fundamentally unreliable for any safety critical purposes, particularly its history of helicopter tragedies resulting from its use in helicopter drive trains.

Keywords: metal casting, entrainment, oxide film, bifilm, defect, fracture

1. Introduction

I have written about casting processes extensively, some would say, over-extensively, but useful new concepts enshrining the new insights into this important subject are being only slowly realized and understood. Their awful importance to the reliability of engineering is yet to be fully adopted. It is useful therefore to offer a short summary in this appropriate volume. A personal reading list is appended.

Space unfortunately forbids any lengthy discussion of the influence of the melting processes on the quality of the resulting castings. However, of course, the quality of the metal prior to casting is of critical importance; a liquid metal, especially if a secondary metal (such as recycled metal, as is common in the aluminum industry) can be literally crammed with defects, so that despite the excellence of the casting process and the heat treatment process, the resulting casting predictably fails all mechanical property requirements and is scrapped. We shall have time to give only a brief mention of this important problem. Our subject is the casting process.

2. Fundamental issues

Castings have had a poor reputation as a result of their poor and variable properties. For many years this was thought to be somehow associated with the turbulence of the pouring process, but the details were not understood, and efforts to control turbulence, despite all claims to the contrary, were failures. No-one was aware of the degree of failure to control turbulence during mold filling because, of course, molds made or sand or steel were opaque: the awful internal damaging mechanisms were unseen and unsuspected.

The breakthrough in understanding came from X-ray video studies of mold filling. Although occasional demonstrations of this technique had been made a number of times over the years, it was only in the 1990s that intensive and systematic studies were carried out at Birmingham University, UK [1]. It was quickly realized that because the liquid metal practically always exhibited a surface oxide film, the mutual impingement of drops and splashes, or the folding of the liquid surface, occurred as oxide film to oxide film. The liquid metal, in general, never made contact with itself. Furthermore, the upper surface of the film in contact with the air was dry. Thus, the mutual impingement processes occurring during turbulence of the surface occurred as dry-film-to-dry-film (**Figure 1**). No bonding occurred between these two ceramic films which for many metals and alloys, including steels, consisted of alumina (Al_2O_3) and similar very stable high melting point oxides.

The practical result of this impingement of two unbonded ceramic films, is the effective creation of a crack in the liquid. This defect is called a bifilm. The turbulent pouring of a liquid into a mold can fill a liquid with cracks. The properties of the subsequent casting are, of course, significantly impaired. This is the fundamental problem of all casting processes. It affects nearly all processes in a major way. It is an issue which cannot be ignored.

Throughout this chapter, it should be kept in mind that if oxides in metals are mentioned, it necessarily means double oxides, in other words, bifilms, which implies cracks. Careful consideration of the entrainment mechanism will convince the reader that the surface oxide cannot be entrained and submerged without it occurring as a doubled oxide to create a bifilm crack; all oxides indicate the presence of cracks in the metal. As will be discussed in detail, the bifilm cracks survive plastic working, and so enter the world of the metallurgist and engineer. Because nearly all our engineering metals are intrinsically ductile, all cracks observed in metals almost certainly originate from the turbulence of the casting process.

The presence of bifilms in most metals comes to the rescue of the reasons why metals fail by cracking. After extensively surveying the metallurgical and fracture literature it was a tremendous surprise to this author arrive at the realization that there was no metallurgical mechanism to explain fracture. The lattice mechanisms such as the dislocation pile-up leading to the initiation of a crack were widely believed but have over recent years seen to be in error. Thousands of pile-ups have

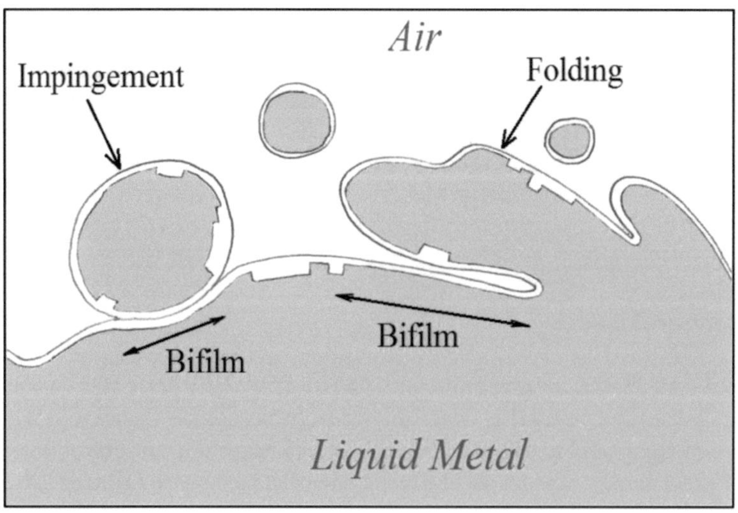

Figure 1.
The impingement and folding processes to produce bifilms as cracks in the liquid metal.

been observed by electron microscopy and studied in detail by computer simulation, but a crack from a pile-up has never been reported. Other theories such as the condensation of vacancies has been known for many years to result not in cracks but in totally collapsed lattice features such as dislocation rings and stacking fault tetrahedra, depending on the stacking fault energy. In brief, the bonds between atoms are simply too strong. Atoms cannot be separated mechanically by any normal forces; pores and cracks cannot be opened up by atomic or lattice mechanisms [2].

The fact that fracture occurs in so many ways and often at modest stresses cannot be explained by conventional metallurgy. This amazing fact is, however, obvious when it is realized that bifilms are present in most metals, usually as a result of poor casting techniques. It follows that if bifilms could be eliminated from metals, there would be no residual mechanism for fracture. Failure by fracture could not occur. This was a sobering realization to this author which it is hoped the reader will be convinced by this short account. If the short account fails to convince, the references at the end of this chapter are recommended.

Before moving on to the discussion of the techniques of casting processes, in addition to the bifilm, a further serious entrainment defect must be described.

In the maelstrom of pouring processes, in addition to the entrainment of oxide films as bifilm cracks, bubbles of air can also be entrained. The bubbles are serious defects in themselves, but their buoyant flotation makes a bad situation worse. Their buoyancy force causes the oxide film at the crown of the bubble to tear, so that it moves to one side, but is immediately replaced by fresh oxide film (**Figure 2**). It can

Figure 2.
Bubbles and bubble trails as collapsed oxide tubes.

be seen therefore that the skin of the bubble effectively slides around the bubble, coming together underneath to form a kind of collapsed tube, which extends back to where the bubble was effectively tethered, the point where it first entered the liquid; probably some early location in the channels of the filling system. This bubble trail is a kind of long bifilm. It can be metres long. Thus, bubbles can create macroscopic crack-like defects out of all proportion to the original size of the bubble. Furthermore, it is common for hundreds or thousands of bubbles to make their way up through the metal, creating masses of tangled defects [1, 3].

The reader may by now be already appalled, realizing the reality of grossly poor metallurgical processing which still bedevils our casting world today. The fact is that as a result of these fundamental entrainment mechanisms, most casting processes are bad. Books are full of the descriptions of casting processes, but none state that nearly all of them usually are capable of delivering only badly defective products.

This short summary will attempt to redress this key issue, illustrating how engineering and the world copes at this time simply by accepting the mediocre properties of metals, often by building in substantial safety factors. For the future, impressive improvements in properties and reliability are forecast for fundamentally improved casting technology.

3. Gravity casting

If liquid metal is allowed to fall under gravity, after a fall of about 10 mm the metal has accelerated to near 0.5 m/s. This is the critical velocity at which the liquid now has sufficient energy to jump or splash up to about 10 mm high, and so be in danger of entraining its own oxide skin during its fall back under gravity. Thus, fall heights and speeds less than these values are safe from the introduction of damage due to surface turbulence. Above these heights and speeds, entrainment of air and oxides becomes increasingly severe [1]. Therefore, when pouring an average sand casting, which might be 500 mm tall, the falling stream reaches speeds of over 3 m/s, far higher than is wanted, so that, in general, copious amounts of defects are entrained. The situation is worse still for the pouring of steel ingots where a fall of 3 or more meters creates speeds of near 8 m/s, generating conditions similar to emulsification with air and oxides.

The skill in the filling of shaped castings by gravity pouring is to limit air ingress into the filling system and limit the velocity at which the metal enters the mold cavity. Only in the last few years have these problems been solved for the first time [3].

The sand casting process (of which there are very many variants) and investment casting processes similarly require these new solutions for design of filling system if, as is usual, filling is by pouring under gravity. Interestingly, these processes both exhibit rather low properties compared to castings poured in metal molds. The improved properties of faster cooled metals are traditionally attributed to a refinement of the dendrite arm spacing (DAS). In steels and Mg alloys there is some truth in this as a result of their limited number of slip planes. However, for Al alloys, with its extremely ductile face centered cubic (FCC) structure, the benefit from DAS is negligible.

The benefit to the faster freezing of Al alloys is a bifilm mechanism. Bifilms arrive in the mold in a compact raveled state because of the dramatically vicious bulk turbulence (high Reynolds number) in the filling system, so that their crack-like morphology is initially suppressed to some extent. Metal molds solidify quickly and freeze in these favorably compact and convoluted defects. In comparison, slow solidification in sand and investment molds allows more time for the bifilms to unfurl. This opening-out process, in which the crumpled bifilms unfold and straighten, resembling the opening of a flower, in which the petals adopt the morphology of planar engineering cracks. The unfurling process generally takes several minutes, and is driven by a number of mechanisms, including gas in solution which precipitates into the 'air-gap' inside the double film, or because of dendrite pushing and other factors [1, 3]. When all the bifilms have straightened out to resemble engineering cracks, the metal properties are at an all-time low. The metal now contains a snow-storm of cracks.

Turning to steelmaking, the technology of casting includes some astonishingly retrograde techniques. In an electric arc furnace, the steel quality is probably quite good as a result of the length of time available for the flotation of oxides. However, the metal quality is ruined by the tilting of the furnace and the fall of metal by several meters into a ladle. The turbulent churning of the steel has to be seen to be believed. However, it takes several minutes for the ladle to be lifted from the pit and taken to the casting station, during which time its quality recovers somewhat because of the very different density of the oxides compared to the dense liquid metal. But this improvement is destroyed a second time by ingot casting. Although some of the damage during casting floats out, not all escapes. The ingot is permanently degraded.

The move to ladle metallurgy is a valuable modern step in steelmaking, but the final pour into the ingot mold is unchanged and undoes much of the good achieved in the ladle.

This problem is especially acute for the casting of special steels, in which the tonnage is often too low to consider the use of the rather superior continuous casting process. Special steels are therefore mainly cast as ingots. At the time of writing, this is a poor process, in which steels which may be required to be especially good for a special purpose are actually made especially badly.

All castings which are top poured under gravity, including many sand castings, nearly all investment castings, and nearly all ingots, suffer the maximum damage from entrainment of air and oxides (**Figure 3**). All top pouring is bad.

In an effort to upgrade the ingot casting process, a bottom gating (sometimes known as uphill teeming) is carried out (**Figure 3**). The reduced splashing by uphill

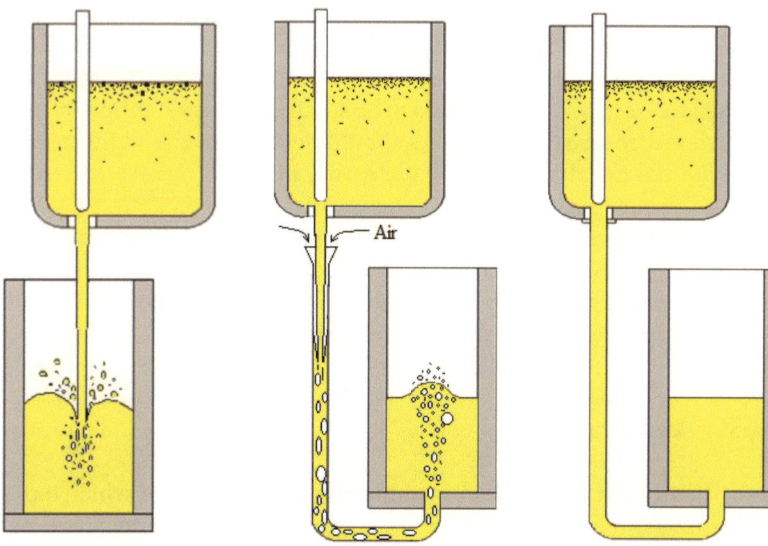

Figure 3.
Illustrating top pouring; uphill teeming; and contact pouring of steels.

teeming improves the surface finish of the ingot. However, unfortunately, the interior quality of the steel is little improved. The falling stream jetting from the base of the ladle enters the start of the filling system at conical intake (often known as the trumpet). The trumpet and following channels need to be oversized with respect to the falling jet to avoid back-filling and over-flowing. This geometry results in at least 50 per cent of the fluid entering the conical basin as air. In the filling system pipe-work, the 50/50 air/steel mix is substantially thrashed together at speeds of up to 10 m/s, ensuring that the bifilm mix will never properly de-segregate, and the bubble trails will further contribute to the copious residual inclusion population, each trail contributing an impressively long crack.

At the high temperatures of some steels, and because of the compositions of some oxides, the crack can evolve to reduce its surface energy. The double film coarsens by diffusion, finally forming sheets of granular solid particles of oxide. The final product is therefore sometimes oxide fragments attached to a void, or gas-filled cavity such as an argon bubble, the residue of the bifilm 'air gap'. The argon bubble remains after the oxygen and nitrogen have been taken into solution in this energetic mixing, leaving the 1 per cent argon in the air as the insoluble residue.

The overall result is that the internal quality of the bottom gated ingot is hardly any better than the top poured ingot.

A dramatic improvement to gravity pouring is achieved by contact pouring (**Figure 3**). The author now insists on contact pouring for all his shaped castings of any metal. The foundries which use this technique find that their cast products are transformed, including cast steels, Ni alloys, Al alloys and bronzes.

Returning to the casting of bulk steels, the continuous casting process certainly delivers a superior product to those steels cast as ingots. This is partly because the ladle take time to be delivered to the top of the casting machine, and then only slowly releases its melt from the base of the ladle – the steel at the base of the ladle having the best quality as a result of the melt cleaning automatically by flotation, and the extended time which is available for flotation, which can easily be 10 times longer than the time required to cast an ingot.

The continuous casting process could probably be much improved by paying attention to important details. The use of tapered nozzles for ladles and launders (the tapering avoids air entrainment into the nozzle which is probably the reason that nozzles block by oxide accumulation [1, 2]). Any fall exceeding 10 mm has to be controlled, so as not to occur in air but submerged under metal or slag. There is a huge amount of research concentrating on the *detrainment* of inclusions from launders, when the research really needs to be spent on the prevention of *entrainment* of inclusions because the inclusions should never be in the launders in the first place. The fall of the metal into the initially empty mold is a massive retrograde step which has to be eliminated – the piling of scrap metal into the mold is a poor starting technique not helping at all. The initial fall creates masses of bifilms which then pollute the whole length of the cast strand because of the progressive dilution of the initially badly damaged metal [2]. These are all simple, negligible cost techniques for which there is no excuse for not implementing immediately.

4. Counter-gravity casting

All the difficulties of mold filling by pouring under gravity, at which metals are accelerated to unwanted high speeds, and so creating masses of unwanted defects, are avoided by not employing gravity.

If now, by some means, the metal can be pumped uphill into a mold, its velocity can be controlled at every point, and need never exceed the critical velocity 0.5 m/s at which entrainment becomes possible. Furthermore, air need never be entrained, so that bubble damage from bubble trails cannot occur. The contrast between conventional gravity pouring and counter-gravity filling is seen in **Figure 4**. In the counter-gravity process the surface oxide film is never entrained; as the metal rises, the surface film simply splits and moves to one side, but instantly reforms and splits, moving aside etc. The surface film becomes the skin of the casting. It is never entrained. In principle, the counter-gravity casting of metals promises perfection.

However, attempts to achieve this perfection are, unfortunately, often not conspicuously successful.

The most disappointing process which nominally adopts counter-gravity filling is the low pressure permanent mold process for the casting of automotive castings, particularly wheels. Most embodiments of this process currently employ a large melting furnace to tip metal into a ladle, in which it falls at least a meter. This damaged metal is then driven by forklift truck to a treatment station, then to the furnace of the casting unit, into which it is tipped again, falling another meter and suffering more damage. The consequence is a really poor quality of metal, full of bifilm cracks, giving poor strength and toughness. If this were not bad enough, there is worse to come! The furnace is pressurized to displace the metal up the riser tube and into the mold (**Figure 5(i)**). After solidification of the casting the release of the pressure causes the melt to fall down the riser tube, thereby displacing all the oxide sediment, which has taken its time to settle at the bottom of the furnace, back into suspension, just in time for the next casting to be made. In addition, the depressurizing action causes bubbles to expand from pressurized gas trapped in crevices in the refractory walls, and the creating of generous quantities of bubble trails. Sufficient bubble trails can sometimes be created to make the metal uncastable; the furnace becomes filled with a slurry of metal and oxide films resembling concrete. Crucible furnaces (**Figure 5(ii)**) appear to be somewhat more resistant to the worst excesses of this problem because of the finer pore sizes from use of isostatic consolidation during their manufacture.

A more recent development is the application of pressure to the mold, pressurizing the incoming metal, and therefore acting to keep bifilms closed, with a benefit to properties. Naturally, this pressure effectively acts to counter the pressure used to pressurize the metal up the rise tube, hence the name 'Counter-Pressure Casting.' However, if counter-gravity is employed to cast good quality metal, in which the bifilm population has been reduced or eliminated prior to casting, the counter-pressure becomes redundant. The counter-gravity counter-pressure process seems

Figure 4.
Conventional gravity casting and counter-gravity casting.

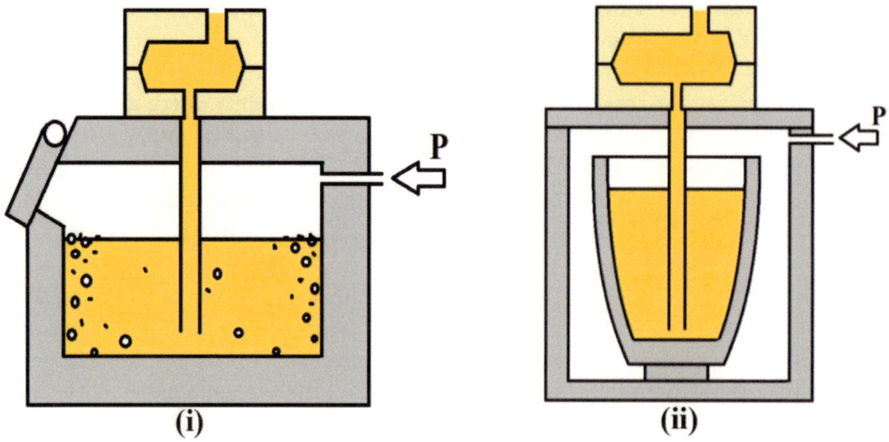

Figure 5.
Low-pressure casting in (i) a refractory lined pressurized furnace, compared to (ii) a pressurized crucible furnace.

to this author to be a step too far. Liquid metals, like all liquids, is effectively incompressible, and cannot be improved by pressure.

When counter-gravity casting is carried out well, with cleaned metal free from dense populations of bifilms, and when transferred uphill, against gravity, carefully controlled by a pump, the resulting castings can be spectacularly excellent.

In his early days in the casting industry, when the author first set up the Cosworth counter-gravity process, the castings requiring aerospace quality were cast in the half of the foundry containing the counter-gravity system using an electromagnetic pump for the liquid aluminum alloy. The other half of the foundry was retained for less important gravity cast products. Eventually however, it was found that with counter-gravity it was difficult to make a bad casting, whereas with gravity casting it was difficult to make a good casting. After 6 months, the gravity area was closed, and all castings were made on the pump.

5. Metal injection casting

The production of castings by high pressure die casting (HPDC) are generally limited to the low melting point metals Al, Mg, Zn and Pb. Some brasses are cast by this technique but attempts to cast stainless steels seem to have been abandoned. This brief description will concentrate only on the casting of Al alloys.

Although the term 'high pressure' seems to offer reassurance of a well-consolidated pore-free product, as most readers will be aware, this can be far from the truth and should never be forgotten by potential users. In general, the HPDC process can never guarantee freedom from porosity and leakage. Nevertheless, the process has valuable features and capabilities which distinguish it widely from other casting methods.

The process is often described as high productivity. It is true it benefits enormously from its ability to cast thin sections which can freeze quickly. But in common with all metal mold casting processes, the metal mold cannot be opened until the casting has frozen, or nearly frozen. This waiting time for the casting to solidify is a major contribution to the production cycle. High production sand casting systems can be much faster for any thickness of casting section, because after pouring, the mold can be moved away, allowing the immediate pouring of a second mold, and so on. Both sand and die systems can benefit from multiple impressions, giving multiple castings per filling.

For most HPDC machines, metal is spooned from an open holding furnace, and poured into a shot sleeve, from where it is rammed into a steel die by a piston. The steel die is sunk into a massive steel bolster, which is kept closed during the shock of the filling process by hydraulic rams developing hundreds or thousands of tons of force. This brutal description is not too far from reality, although the injection stroke and filling pattern is now often optimized by computer simulation to reduce air entrainment, which has resulted in significant improvements to the reduction in porosity in castings.

The turbulence during the injection process, in which the metal velocity usually exceeds 50 to 100 m/s, is so great that defects are necessarily created but are accepted as a feature of the process. Interestingly, the high density of bifilms is not necessarily the disadvantage that might be imagined; the long oxide flow tubes (the oxide tubes which surrounded the jets of metal entering the mold cavity) and other bifilms are aligned along the flow direction, giving a fibrous microstructure whose properties somewhat resemble the directional features of wood. The rapidity of the filling process, being completed within milliseconds, probably also suppresses the degradation of the casting by bifilms, whose constituent films have so little time to grow and are necessarily extremely thin. Their limited thickness may permit some bonding between the two films as a result of atomic rearrangements during their transformation from pure alumina to spinel as Mg in the alloy diffuses into the bifilm. The high pressure, keeping the two sides of the bifilm closely in contact is a further aid to bonding and, in any case, provides strength by the bifilm being enabled to resist shear force, because of jogs and wrinkles, if not direct tensile force. Even so, the HPDC castings can never be relied on not to leak, and sometimes, not to fail unexpectedly. Their use for safety critical purposes should therefore only be accepted with very great caution. (In contrast, gravity sand and gravity die castings [permanent mold castings] are typically favored for safety critical components).

Traditionally, small HPDC machines provide high productivity for small thin-walled products. The accuracy and surface finish are good, often eliminating machining, making the process favored by engineers. Recently, extremely large HPDC machines have been built to produce castings of several square meters area with walls only a millimeter or two in thickness, creating large pieces of automobiles in one shot.

6. Vacuum casting

There are some genuine reasons why vacuum is needed for the melting and casting of certain alloys and certain products. Sometimes, a limit on the oxidation of reactive metals or alloying elements is required. At other times the vacuum is needed to ensure the filling of extremely narrow and tapering sections as in turbine blades.

Alternatively, vacuum casting is used, imagining that this will prevent the formation of defects during a top pour. This appears to be a widespread but dangerously incorrect assumption. The entrainment defects resulting in bifilm creation appear to be the same no matter what environment is used, whether this is air, inert gas or vacuum. The reason is that both the inert gas and the vacuum environments always contain sufficient oxygen and/or nitrogen to create oxide or nitride films on the surface of the pouring liquid, so that defects of identical size and geometry are formed if entrainment of the surface occurs – the only difference being the thickness of the resulting bifilms. Bifilms are generally so thin that they are not easily seen when cast in air, but are, of course, far more difficult to detect in vacuum castings. The vacuum casting has a lower oxygen content, and is assumed to be cleaner, which in a way it is. But the distribution and sizes of its population of cracks appears to be unchanged [2].

7. Vacuum induction melting (VIM)

The formation of bifilms in vacuum casting is practically universal, because ingots and castings poured in a vacuum furnace are nearly always top poured. In huge industrial VIM installations, the fall can be many meters, creating much damage to the metal. For instance, all the metal used by the aerospace industry for remelting for the casting of turbine blades is damaged during VIM preparation of the Ni-base alloys; the metal is top poured, falling many meters, down long vertical steel tubes; the larger the diameter of the tubes the worse the damage to the alloy by splashing and entrainment.

In probably all the leading R&D institutions in the world, metals and alloys for research are melted and poured in laboratory VIM furnaces, the top pouring, with the metal falling by a meter or more, fundamentally undermining or complicating nearly all metallurgical R&D worldwide (**Figure 6**). It has greatly contributed to the lack of understanding of more complex failure forms of metals such as fatigue, stress corrosion cracking and hydrogen embrittlement among others as a result of all researchers being unaware that their research materials were densely pre-cracked [2].

It is with great regret therefore that we have to conclude that the preparation of most metals and alloys by vacuum casting is a snare and delusion. It would be easily possible to make castings in air of far greater perfection by simply avoiding surface turbulence during the casting process. This is most effectively achieved by

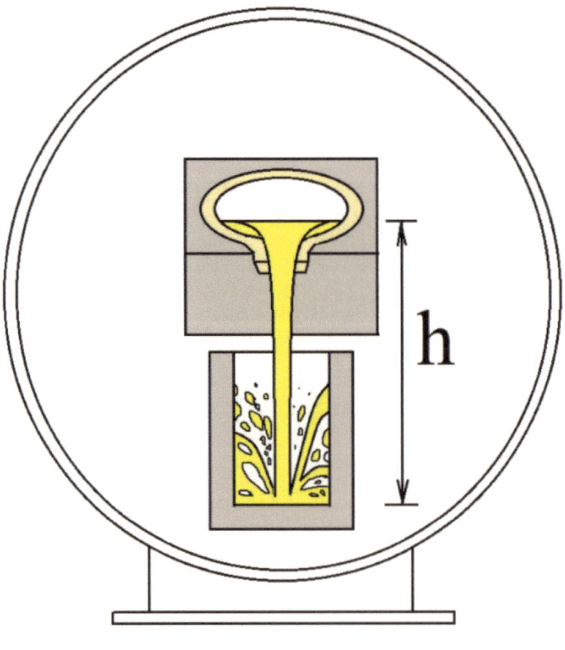

Figure 6.
A simple laboratory vacuum induction furnace illustrating the awful top pouring, creating damaged products.

abandoning gravity pouring and adopting counter-gravity filling of the mold. The world needs to convert its casting operations to counter-gravity casting. The suffering of the casting world from the ubiquity of casting defects will then be of interest only for historians.

8. Secondary remelting

The secondary remelting processes for steels and Ni alloys are designed to deliver a premium quality of metal in the form of an ingot. Their starting material is a reasonably good metal in the form of a consumable electrode which is slowly and progressively remelted by arc, plasma, electron beam, or joule heating in a liquid slag layer etc. As the tip of the electrode melts, a new ingot is then slowly built up drop by drop within its protective environment of vacuum or slag. The ingot solidifies tolerably rapidly because of the use of a water-cooled mold.

At the time of writing, it requires to be noted, with regret, that none of the secondary remelting processes are totally reliable. All can have serious crack defects which can survive the subsequent forging or rolling, and the heat treatment, making these products unreliable in service. Some, as we shall see, can be seriously unreliable.

9. Vacuum arc remelting (VAR)

VAR is probably the most widely used of all the secondary remelting processes (**Figure 7**). The marketing of VAR benefits from its name: engineers are attracted to the concept of 'vacuum' suggesting cleanness.

However, the VAR process is particularly susceptible to its slightly oxidizing vacuum conditions, growing an oxide skin on the horizontal ledges formed by the slow layer-by-layer advance of the solidifying liquid. This variety of advance occurs because of the strength of the oxide on the advancing meniscus as it rolls over the solidified or solidifying metal around the edge of the ingot. The vertical advance occurs by the horizontal flow of the liquid front, gradually spiraling upwards, advancing vertically by the 8 mm high steps corresponding to the height of the meniscus. This is the height which surface tension can support against the hydrostatic pressure due to this depth [1]. As the meniscus rolls over the oxide film on the

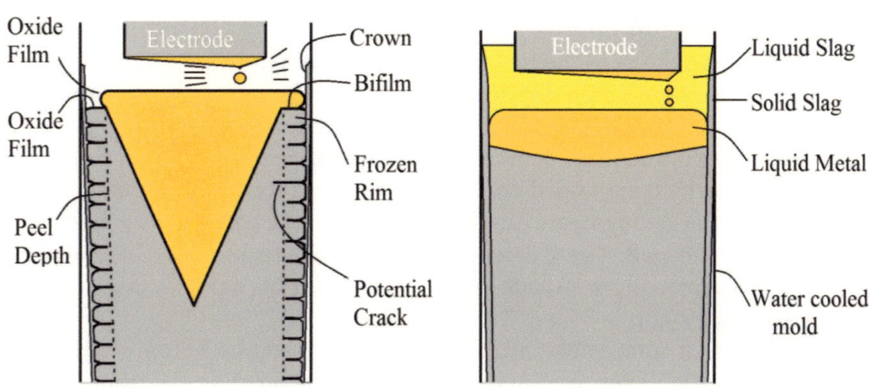

Figure 7.
VAR and ESR secondary remelting processes.

freezing ingot, the meniscus lays down its own oxide film on top of the surface oxide film, creating a bifilm. It is a substantial crack, possibly extending up to 50 mm deep [2].

The presence of cracks around the circumference of VAR ingots is widely known. It is proven by the cracking of the ingots in forging (in contrast to ESR ingots which forge like butter). The manufacturers machine off around 5 mm depth of the outer surface as a token gesture to remove cracks. Because ingots forge better after the removal of the 5 mm it is certain that most cracks are removed. However, of course, it is unlikely, given the variability of conditions during arc melting, that all will have been removed.

The falling-in of the 'crown' of spatter and evaporated metal (**Figure** 7) into the forming ingot may introduce additional macroscopic bifilms. A further source of major bifilms is the electrode. The electrode is typically made by top pouring into an ingot mold, sometimes in air and sometimes in vacuum (the VIM/VAR process combination) but as we have seen, whether air or vacuum, the seriously deleterious defect distribution will be essentially the same. A bifilm taking up a substantial area of the cross section of the electrode may cause a large piece of the electrode to detach and fall into the melt. This unmelted fragment will be effectively surrounded by a bifilm (its own oxide surface collecting a covering of the oxide on the liquid as it plunged through the surface) together with its own internal oxide bifilms.

10. Electroslag remelting (ESR)

Nearly all producers of VAR material also produce ESR. The ESR process is probably the next most popular secondary remelting process. The tip of the electrode is heated by the passage of an electrical current through a slag layer (**Figure** 7). The thin film of melted metal, gathering over the base of the electrode, and finally detaching and falling through the slag droplets of metal, ensures that metal arriving in the melted pool contains only rather small bifilms. Any bifilm which happens to touch the slag will be sucked out of the liquid metal and into the liquid slag by capillary attraction: the solid oxide will be wetted by the slag, a mainly oxide liquid. It will then be dissolved in the slag and disappear. This sets a limit of around 1 mm for the maximum size of bifilm defect which could be present in an ESR ingot (contrasting interestingly with the potential for 50 mm defects in VAR). This ability of ESR to actively extract oxides and dissolve them is a fundamental and unique benefit of the ESR system. (In the early days of the process, no-one could understand how the ESR process improved the properties because no significant changes to the metallurgical structure could be seen!).

There remains a threat to the integrity of ESR material through no fault of the ESR process itself. The threat lies once again with the desire to provide only the cheapest electrode, and so, once again, electrodes are usually cast by top pouring. An electrode top-poured in air contains bifilm defects as surface-appearing laps which can be seen by the unaided eye from 100 m distance. It is no wonder therefore that, once again, large fragments can detach from the electrode during melting and can fall into the melt. These defects contain unmelted and unrefined material. The author has personally seen such a defect the size of his hand on the section of a 600 mm diameter ingot.

For the future, if completely reliable metal is required, the ESR process is the only currently available source, but requires the provision of an electrode cast by a reliable process. Such a process includes low-cost ingots cast by contact pouring (especially if enhanced by flush filters and spin traps), or perhaps an improved continuously cast material, or, ultimately, counter-gravity casting of some kind.

The world would then have, for the first time ever, a totally reliable metal process free from macroscopic cracks, but containing only microscopic cracks of maximum size perhaps 1 mm.

11. Metal processing

The result of entrainment of bifilms during casting production results in huge losses in metals processing such as forging, rolling and extrusion. All these processes suffer from cracking of the processed metal, sometimes to the extent that the metal cannot be processed. Many steel ingots suffer from cracking during cooling. For this reason, fluted molds assist to disperse stresses across the faces of the ingot, although the technique is not especially successful for some steels. The break-outs of liquid steel from the cast strand during continuous-casting are almost certainly bifilm problems which is the reason this rather common disaster has remained unsolved. A number of Ni-alloy ingots are known for cracking at the first stroke of the forge, and rolled steels suffer edge cracking, longitudinal cracks, transverse cracks, internal cracks. Many metals suffer edge cracking during extrusion and rolling. Aluminum alloy semi-continuously cast slabs suffer cracks of all sorts, some measuring their length in meters across the slab face.

For those working in metal processing, valuable R&D has often been carried out to provide process 'windows' defining the limits of successful processing. Many such limits have been set by the onset of cracking. Processors would be delighted to see these limits eliminated. The processing of metals remains to be revolutionized.

12. Metal properties

If the metal is successful to survive processing, it then can suffer from its internal bifilm population during its service life. All its mechanical properties and failure modes are affected by its bifilms. Some of these aspects are discussed below.

12.1 Ductility

Practically all of our engineering metals are intrinsically ductile. Basic dislocation theory predicts that if a stress is applied to a crack in most engineering metals, dislocations are emitted prior to the advance of the crack tip. The result is that the crack blunts, and crack propagation cannot occur.

(This behavior contrasts with the rather few intrinsically brittle metals, including W, Cr and Be, for which the imposition of a tensile stress causes the crack to propagate first, without the emission of dislocations. Fracture by cleavage is a variety of brittle failure but is only known for certain to exist in zinc).

In theory, therefore, tensile overload in the majority of our metals should result in plastic necking down to 100% reduction in area (RA) despite the metal possibly having high strength, resulting in high stress supported during the plastic failure.

Cast aluminum alloys fail this expectation lamentably, having typical elongations to failure in single figures, typically 3 ± 3%. Al alloys generally contain a dense populations of bifilm cracks because the alumina bifilms are slightly denser than the liquid, but contain some entrained air lending some buoyancy, causing bifilms to be close to neutral buoyancy, and thus remaining in suspension for hours or days. Conversely, steels typically reach 50% elongation because the rapid flotation of bifilms within minutes results in much cleaner metal. Other factors leading to some bonding across the central interfaces of bifilms in some steels further contribute to improvement [2].

In contrast to steels, the lack of a definitive yield point in Al alloys is probably due to the presence of bifilms, raising the stress around the bifilm because of the loss of load supporting area and the sharpness of the bifilm crack. Thus, plastic flow occurs early, spreading from scattered locations throughout the matrix of an Al alloy before the macroscopic yield point is reached. Similarly, the lack of a fatigue limit in Al alloys compared to steels can be similarly explained.

12.2 Fatigue

In the experience of the author, much of the area of many fatigue fracture surfaces is comprised of bifilms. The genuine fatigue areas characterized by 'beach marks' appear to be generally confined to a few regions which happen to be devoid of bifilms. The remainder of the surface is often described as quasi-cleavage failure, which is simply a polite admission of ignorance – no-one seems to know what quasi-cleavage is, except that it is definitely not cleavage. These regions appear to be bifilms, hiding in plain sight. The regions often outline grains because the bifilms tend to be trapped intergranularly between grains or are straightened by dendrite growth transversely across grains. When the advancing fracture reaches the limit of one bifilm and has to migrate out of its plane to continue its advance by opening the next bifilm, the plastic shearing process between bifilms outlines the grains.

A typical well-known example is the fatigue failure of the main bearings of wind turbines. These huge steel rings are forged from a single large ingot. The interior surface of the ring is naturally composed of the center of the ingot. Bifilms will have been segregated here by dendrite pushing. Because of the huge size of the ingot, the plastic deformation involved in forming this into a ring is modest; the bifilm cracks are merely pushed around a little but are by no means 'welded' closed. Large tangled masses of bifilms are therefore present on the inner surface of the bearing ring. These masses of pre-cracked regions are likely to be millimeters or even centimeters across. They experience the high (2000 MPa) rolling stresses, with the result that minute connections inside these regions, or linkages holding the masses to the matrix, will suffer even higher concentrations of stress, resulting in genuine fatigue failure of the tiny isolated connections holding the pre-fractured regions together. Ultimately, whole, macroscopic blocks of material break away among the rollers because of the minute, almost negligible amounts of fatigue, signaling the imminent death of the bearing.

12.3 Creep

There is excellent evidence for creep being significantly controlled by the presence of bifilms. In the comparison between polycrystal and single crystal turbine blades, it was traditionally explained that the overwhelming benefit to resistance to creep failure was the elimination of the transverse grain boundaries. It was assumed that the boundaries were weak. However, as much recent research has now demonstrated, grain boundaries are immensely strong. The traditional explanation is clearly unsatisfactory.

The realization that bifilms are present in the liquid alloy leads to a logical explanation. In the conventional polycrystalline casting grains nucleate and grow randomly throughout the cooling liquid. Bifilms in suspension therefore become trapped as grains collide, the bifilm effectively becoming coincident with the newly formed grain boundary. The boundary is therefore weak, effectively pre-cracked, and the polycrystalline casting is observed to have poor creep properties as a result of a high proportion of its boundaries harboring cracks.

In contrast, in the conditions for growth of the single crystal, the slow vertical advance of the freezing front will push bifilms ahead. Those that are not pushed may float. The result is a casting relatively free from bifilms, and displaying astonishingly good creep life.

12.4 Pitting corrosion

Bifilms can act as invasive pathways for corrodents into the interior of metals. The outside surface of a metal may be tolerably resistant to corrosion, but at the location at which a bifilm emerges, breaking the surface, the ingress of rain or salt water is likely to form an etch pit. The localized corrosion around the bifilm may be enhanced by precipitates of second phases and intermetallics which favor the wetted exterior surface of the bifilm (its wetted exterior surface contrasts with its dry, unbonded inner interfaces). These different compounds with different electrochemical potentials attached to the exterior surface of the bifilm can provide vigorous corrosion couples.

Figure 8 shows a typical etch pit. Although the conventional explanation of the image would be that the etch pit has initiated the formation of cracks, the reverse is true. The cracks are bifilms, as can be identified from their morphology and precipitates. They have initiated the etch pit.

In the past decade there have been at least three, perhaps four or more, helicopter crashes, some extremely tragic, in which items of the drive train appear to have failed by fatigue initiated from an etch pit. Experts from around the world have been puzzled because an etch pit was far too small to have initiated the fatigue crack. In the case of one main rotor shaft, which appeared to have failed in this way, the shaft was designed with a safety factor of five. It is not conceivable that such a robust shaft could be threatened by an etch pit.

It is easily appreciated, that the etch pit is merely the witness to the presence of a bifilm crack. Furthermore, the bifilm could have been extensive, such as possibly extending over a major portion of the shaft. The shaft was formed, of course, from VAR steel, so that the probability of its being pre-cracked is virtually certain. The crack

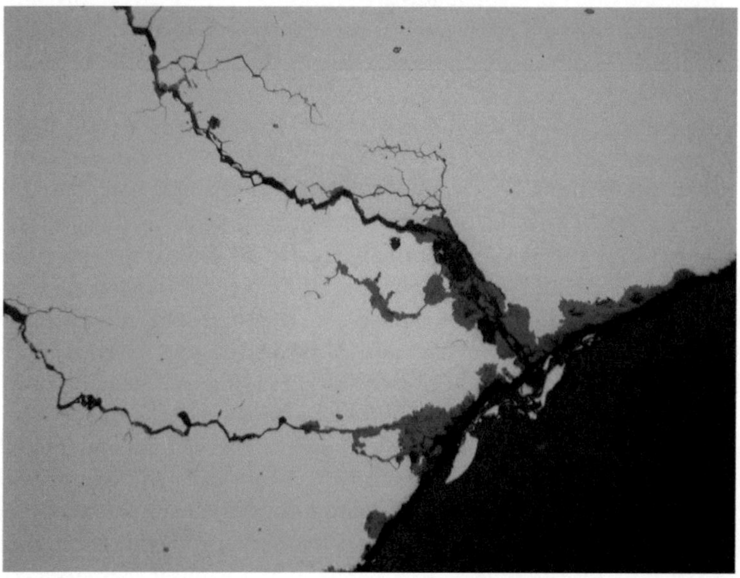

Figure 8.
Etch pit in a steel turbine blade. Courtesy Metallurgical Associates Inc.

would have evaded detection because, being formed by oxidation in vacuum, its oxide films would have been extremely thin. Also, as a universal feature of castings, and heat-treated products, especially if quenched, the interior is in tension, but the exterior surfaces are in compression. The crack on the outside of the shaft would therefore have been tightly closed.

Attempts to find bifilms by nondestructive testing (NDT) has proven to be tragically unreliable. As always in such difficulties, the clear way forward is to use only those processes which do not generate bifilms and which are therefore intrinsically reliable.

12.5 Stress corrosion cracking (SCC)

This dangerous failure mode involves almost no loss of metal by corrosion but can generate deep cracks by a time-dependent advance, often under only low stress. A metal can be sensitized to SCC by heat treatments.

There seems to be good evidence that SCC is a bifilm phenomenon, whereby the corrodent is simply moving through the 'air gaps' of the bifilms, linking bifilms by corrosion, driven by the stress concentration at the bifilm linkages [2].

The action of certain heat treatments to enhance SCC susceptibility is here proposed to arise from the precipitation of second phases on the bifilm. The favored formation of precipitates on bifilms seems to be the result of the reduction in the strain energy of formation, because the volume change and shape change of the new arrival can be more easily accommodated by the 'air gap' of the bifilm. The movement of part of the new phase into the air gap is likely to assist the forcing open of the air gap, so that percolation of the corrodent is facilitated [2].

12.6 Hydrogen embrittlement (HE)

There are numerous theories which have attempted to explain HE, but the phenomenon cannot yet be claimed to be clearly understood. In practice, the ingress of hydrogen into a stressed steel can result in gradual loss of ductility, and final fracture. The process has been identified as the slow progress of a crack until the final fracture when there is insufficient area to support the load. Hydrogen enters the steel as a proton released from certain corrosion mechanisms. For research purposes, hydrogen is introduced by electrochemical processes. Significantly, researchers complain about the interference of blistering which upsets their experiments during the charging processes, and report they are at a loss to know how the blisters can nucleate [2].

Once again, the bifilm seems more than adequate to explain all these observed characteristics of HE. The blisters are the observation of bifilms, inflated by hydrogen, near to the surface of the metal. Clearly, bifilms in the interior of the metal will also be experiencing the pressurization of hydrogen gas. Bifilms will almost certainly aid the progress of the gas into the interior of the metal, greatly accelerating the apparent rate of diffusion. However, it is probably mainly those isolated bifilms whose internal pressurization is leading to internal stress build up which is countering the ability of the metal to withstand tension.

There has been much interest in the attempts to desensitize a metal to HE by providing sinks for hydrogen. The sinks have been generally thought to be dislocations, and stress fields around carbide precipitates. This author has proposed that bifilms are probably significantly more capacious sinks, and the action of carbides is to precipitate onto bifilms and to prize them open, enabling them to accommodate even more hydrogen. He suggests that in an increasing supply of hydrogen, the bifilm would act as a temporary reduction in the deleterious effect of hydrogen, but this benefit would only exist at low hydrogen levels. When the hydrogen pressure in the bifilms equalled and the exceeded the yield stress, the damaging effects of HE would resume unchanged [2].

13. Conclusions

1. The entrainment of the oxide film on the liquid metal during casting processes leads to widespread damage to metals. Pre-cracking by a poor casting technique is central to the loss of properties, and to numerous failure modes, including those during solidification, during cooling to room temperature, during metal processing, and during service conditions.

2. Casting processes involving top pouring are especially damaging.

3. Casting processes using gravity pouring can be designed to yield significant benefits in bifilm reduction and are recommended if counter-gravity cannot be provided [3].

4. Ultimately, counter-gravity casting is strongly recommended to be the new casting norm, capable of delivering defect-free cast products.

5. The current use of VAR steels in all critical applications (especially such applications as helicopter drive chains) appears to be dangerously unreliable.

6. The reliable secondary remelting process could be ESR if combined with a reliable electrode. The implementation of this process combination would be greatly valued by the engineering world.

7. Both primary and secondary casting processes can now be made to deliver economic metals which cannot fail; metals we can trust.

Author details

John Campbell
Department of Metallurgy and Materials, Faculty of Engineering, The University of Birmingham, UK

*Address all correspondence to: jc@campbelltech.co.uk

References

[1] Campbell J. Complete Casting Handbook. 2nd ed. Oxford, UK: Elsevier; 2015

[2] Campbell J. The Mechanisms of Metallurgical Failure—The Origin of Fracture. Oxford, UK: Elsevier; 2020

[3] Campbell J. Mini Casting Handbook. 2nd ed. Malvern, UK: Aspect Design Limited; 2018

Chapter 2

Solidification of Metals and Alloys

Upendra Kumar Mohanty and Hrushikesh Sarangi

Abstract

In order to analyse the process of solidification of metals and alloys critically, it is most pertinent to understand the different modes of nucleation and the uneven rates of growth throughout the melt. It is also important to take a note of the constraints in the growth process that definitely influence the crystal structure and the structure related properties of the casting. The freezing pattern of the liquid melt decides the feeding of the mould which is instrumental in producing a complete and compact casting. For pure metals and even in case of alloys with a narrow freezing range a well defined solid–liquid macro-interface exists. Here feeding of the solidifying casting is the easiest, by the common lowering of the liquid metal surface in the mould. However, in many instances, a well defined interface is not witnessed. The solid–liquid interface could be discrete and not continuous. Here process of feeding the solidification sites that witness considerable shrinkages, may become complicated. On grounds of above it is implied, the process of solidification constitutes an important aspects in the production of a defect free casting.

Keywords: solidification processes, modes of nucleation, rate of growth, crystal structure, freezing pattern solid liquid interface

1. Introduction

The manufacture/fabrication of all men- made objects made up of metals/alloys involve the process of solidification at same stage. It processes of phase change, a liquid phase giving way to a solid phase. In metals and alloys, however, solidification involves the formation of crystals, a crystalise solid exhibiting regularity in atomic spacing over a considerable distance [1]. This is dissimilar from a process involving glasses and polymers. However, even in the case of metal/alloy, when crystals are deposited as a consequence of solidification from the melt, though, there exist certain extents of internal symmetry, there are certain irregular external forms and shapes. This tendency can be attributed to the uneven growth rates throughout the process of solidification and the constraints in the growth process during the last stages of freezing. In the ordinary general case, the solidified grain or cell representing the structure of the solid is a normal crystalline unit formed in the cast structure. However, under the specific case of eutectic freezing, the solidified cell consists of two separate crystal structures with the simultaneous growth of separate phases [2].

The most important practical applications of the process of solidification is found in the production of castings. Indeed, casting can be defined as liquid metal forming. The process consists of introducing the liquid metal of appropriate composition into the mould effecting its solidification under controlled conditions of cooling, pouring, etc. to obtain the desired cast structure [3, 4]. A molten metal has a viscosity which is about one-twentieth of the corresponding solid. Thus, instead of spending, high energy, overcoming the high flow of stresses of a solid to shape if by adopting bulk metal forming operations of forging, extrusion, rolling, etc. a liquid metal with essentially zero shear stress is required to be handle. A detailed study of the process of solidification, therefore, enables one to know and hence control the microstructure of the casting that decides is microstructural properties.

The process of freezing of a solid from its melt is accompanied by to very important phenomena which decides the intrinsic properties of the resultant casting. At the first instance, freezing is associated with volume contractions as a consequence of the development of a more closely packed solid [5]. At the same time a reduction in the molecular motion is experienced when the randomly moving molecules in the liquid phase generate the nuclei that finally grow into the solid phase. Latent heat of crystalization is liberated at the solid/liquid interface. This liberated heat energy markedly affects the rate and mode of crystal growth. The general fall of temperature to give way to freezing causes a lowering of the solubility of the alloying elements in the melt. Solute atoms are rejected at the solid–liquid interface. The solubility of the alloying elements is further affected as a result of the changed composition of the alloy grossly affecting the final structure of the solidified melt.

Where the casting process is the last stage of fabrication or it has to be followed by further mechanical working, solidification possesses pay an important role in deciding the microstructure of the product and hence its final structure related properties. In this respect, two distinct cases can be considered:

1.1 Casting, the final operation in production

One of the major problems concerning the process of casting involves the local variation of the resultant microstructure leading to compositional variations. The above is illustrated in **Figure 1**.

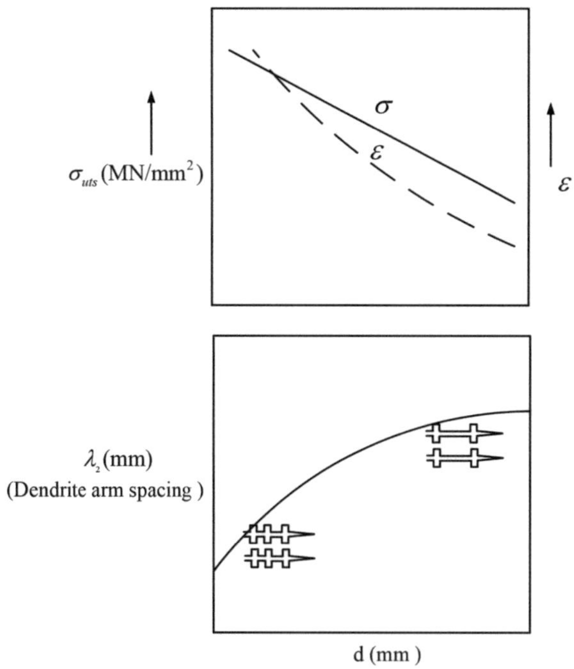

Figure 1.
Schematic presentation of alloy properties as influenced by its position in the melt.

As seen in **Figure 1**, the dendritic arm spacing (λ_2) is measured from the surface of the casting in terms of 'd'. Clearly, with the increase in 'd', i.e., when one goes further into the casting, λ_2 increases. This leads to the resultant variations in the properties of the casting, such as its strength 'σ' and elongation 'ε'. The finer microstructure, in general, exhibit superior mechanical properties. In turn these finer structures are generally formed when the rate of solidification is high. Such high rates of solidification are witnessed as or within a short distance from the surface of the mould. Thus, good combination of mechanical properties are ensured close to the mould surface. As the distance increases from the surface of the mould, into the interior of the casting, rate of solidification decreases, λ_2 increases and grain size of the casting increases rendering a poor combination of mechanical properties.

1.2 Castings with heavy mechanical working

Heavy reduction through mechanical working is not a very efficient method of modifying the edge cast structure. Any initial heterogeneously develop structure, due to the adoption of a faulty solidification process has some tendencies to persist. Therefore, it can be said with authority, any effective control of product quality must be exercised during the process of solidification itself.

The process of solidification comprises of successive stages of **Nucleation** and **Growth**. Whether, freezing is directional or adopts a discrete fashion, throughout the melt, depends on these two factors. In this regard, the location and relative rate of **Nucleation and Growth** hold the keys to the casting characteristics.

2. Concept of nucleation

Any phase change has to get initiated by the emergence of the new phase at some instant of time. Likewise, when a solid phase emerges out of the liquid metal/alloy, it begins with the appearance of Nuclei. These are the cluster of atoms which come together during their course of random motion in the melt and can be termed as the embryonic crystals. These embryonic crystals permit further sitting at atoms on their surfaces which causes the growth of the solid phase. However, many of the nuclei again disappear in the melt, the clustered atoms again moving randomly in the melt. Only those nuclei which are stable and meet the thermo-dynamic requirements, allow growth to take place on their surfaces.

Nucleation can take place in two ways.

2.1 Homogeneous nucleation

The coming together of the randomly moving atoms, from within the melt to form the embryonic crystals, the Nuclei, is known as Homogeneous Nucleation. These are smaller zones of higher density, formed by the ordered cluster of atoms [6]. Mahata et al. have conducted experiments to understand homogeneous nucleation in solidification of aluminium by molecular dynamics simulation [7]. They are of the opinion that there are many methods like X-ray scattering [8] etc., to monitor solid to liquid transformation. However, these methods are limited by several factors that make it difficult to study Homogeneous Nucleation in pure metals.

The precipitation of the group of atoms as the embryonic fresh phase in the melt, is subjected to a change in the free energy. The total free energy change comprises of two components, VOLUME FREE ENERGY CHANGE and INTERFACE FREE ENERGY CHANGE. Thus, a thermodynamic set of conditions is set up for the formation of the Nuclei and for the Nuclei to be stable and not to dry–out prematurely, these thermodynamic conditions have to be met with.

2.1.1 Volume free energy change

Thermodynamically, when a solid comes out as a liquid, there is a negative free energy change in the system. This change of free energy is directly proportional to the new volume(solid) transformed. Thus, for a spherical solid particle formed in a liquid,

$\Delta G(Volume) = -\frac{4}{3}\prod r^3\,\Delta G_v$, where $\Delta G(Volume)$ - the change in free energy due to the formation of the new volume.

$'r'$ - the radius of the freshly created spherical solid and ΔG_v - the bulk free- energy change per unit volume of the spherical solid created.

2.1.2 Interface free energy change

Thermodynamically, when a new interface is generated due to the emergence of a solid from a liquid, there is a gain of free energy at the interface created. This free energy gained is gained is proportional to the surface area of the solid particle created. For the same sphere, as considered above, with a radius of $'r'$, the free energy gain can be given as:

ΔG (interface) = $4\prod r^2\gamma$, where, all the terms have their usual meaning, $'\gamma'$ being the interfacial free energy per unit area of a spherical surface.

The volume free energy change and the interfacial free energy change are both presented graphically in **Figure 2**. The **Figure 2** also depicts the overall free energy change as a consequence of the two components when the solid volume is created in the melt.

As seen in the **Figure 2**, for small values of $'r'$, the sum of free energy changes is positive. However, as $'r'$ increases, this sum becomes negative. The peak positive value

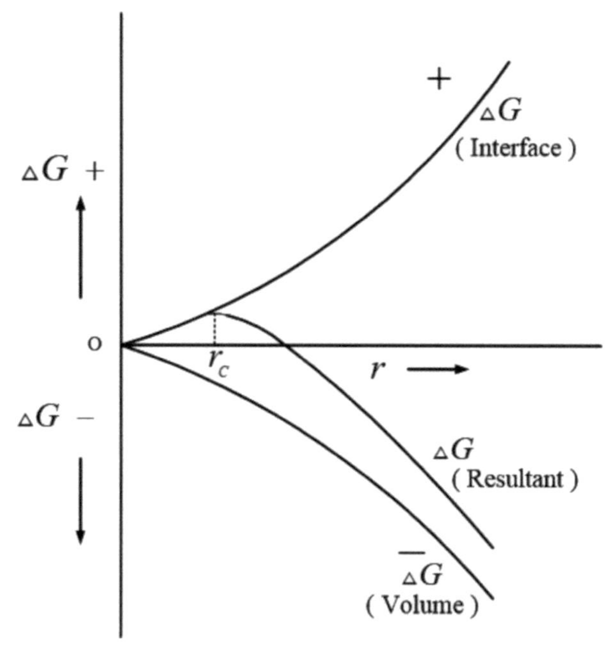

Figure 2.
Change of free energy (volume and interface) as a consequence of the creation of solid phase.

corresponds to the critical radius 'r_c' or the embryonic crystal. The radius of the embryonic crystal must be greater than 'r_c' so that the free energy change ΔG becomes negative, the embryonic crystal becomes stable and growth proceeds. On the other hand, till the 'r_c' is reached, the sum of free energy change remains positive and creates a barrier, obstructing nucleation and the consequent growth. An intent analysis reveals as the temperature falls 'r_c' goes on decreasing. This means, with the fall of temperature, more and more embryonic crystals tend to be stable and the probability of homogeneous nucleation is increased, permitting the growth process to proceed.

From the above it follows, homogeneous nucleation conditions are not favourable at the beginning for the stability of the nuclei as considerable undercooling is necessary for homogeneous nucleation to be effective. In the practical case of casting in a foundry, however, the melt need not be supercooled to make the homogeneous, stable nuclei form to start the solidification process. This is because, in the practical melt in the foundry, solidification processes are initiated by heterogeneous nucleation.

2.2 Heterogeneous nucleation

For heterogeneous nucleation, the initial interface for growth is provided by a foreign particle [9]. This foreign particle can be provided from outside or formed in the melt itself. The impurities, foreign particles or even the mould wall (the substate) can provide for a part of the surface energy required for nucleation. It is a known fact that less activation energy (free energy barrier) is required for nucleation. Therfore, the presence of the substate as mentioned above reduced the free energy barrier and can be very helpful in creating more growth capable nuclei. This is known as heterogeneous nucleation which need less activation energy than homogeneous nucleation [10]. This second phase to act as a nucleus, however, must be capable of being wetted by the melt forming low contact angles and also it must have some structural affinity with the crystalline solid to be formed on it. This second phase could be any one or any combination of the following:

a. impurities in the metal

b. the wall of the mould

c. deliberately added particles to encourage a particular mode of crystallisation

Once the heterogeneous nuclei meet the growth conditions, growth occurs on them. After a certain lapse of time, when the temperature of the melt is lowered, the homogeneous nuclei become stable, and more solid may get deposited on them. At the same time, fresh nucleation may occur generating further stable nuclei. These fresh nuclei may be of the same phase as the first nuclei or of a different phase.

3. Growth processes

The growth process is conceived as the sitting of further atoms on the stable nuclei which brings in the growth of individual crystal or a general growth in the mass of the solid as solidification proceeds *the latent heat of crystallisation* is liberated at the solid–liquid interface. *Zones of thermal supercooling* are generated in the liquid pool. Also, with the lowering of temperature the solubility of an alloying element in the liquid melt decreases. As a consequence, the solute is rejected at the solid–liquid interface. The equilibrium freezing temperature of the alloy is continuously altered and a phenomenon known as *constitutional supercooling*, takes place. Both thermal and constitutional supercooling obstruct growth and alter the growth pattern.

3.1 Plane-front growth

Nucleation does not take place randomly in the metal/alloy melt throughout the liquid because in the actual case of solidification, and uniform lowering of temperature throughout the melt cannot be obtained. There exist a thermal gradient between the cool mould wall surface exposed to the ambience and the interior of the solidifying melt that would eventually form the casting. Therefore, in the practical case, nucleation is initiated at the mould surface and the growth of the solid phase proceed being directed towards the centre of the casting. This growth takes place in a preferred crystallographic direction as dictated by the characteristic of the solidifying crystal. For an example, in a cubic crystal the preferred crystallographic direction is <001>. With the aid of the temperature gradient, the grains oriented favourably grow at a faster rate than the others.

With the lapse of time, depending on the no of effective nuclei and the initial growth rate setup by the initial temperature gradient, the growth of the crystals in the lateral direction gets obstructed. This is because the laterally growing crystals impinge into each other restricting growth of the neighbouring crystals. Also any growth of any crystal ahead of the others, into the high temperature melt is inhibited due to the unfavourable temperature conditions. Such a situation gives rise to planar or plane front growth where in a seemingly plane interface proceeds into the melt causing growth [11]. The interface is plane macroscopically whereas in actual, it is a terraced structure microscopically. This is a typical condition leading to the formation of columnar grains which is often observed in cast ingots. These columnar grains grow in a direction opposite to the direction of heat-flow. This is illustrated in **Figure 3**.

The occurrence of planar growth giving rise to a columnar structure involves thermal condition present in **Figure 4**.

It is assumed, a positive temperature gradient exists at the solid–liquid interface. Here, the liberated latent heat of crystallisation is not enough to reverse the temperature gradient due to the freezing; i.e., a situation is not created when some pockets in the inside of the melt, away from the interface, are at relatively lower

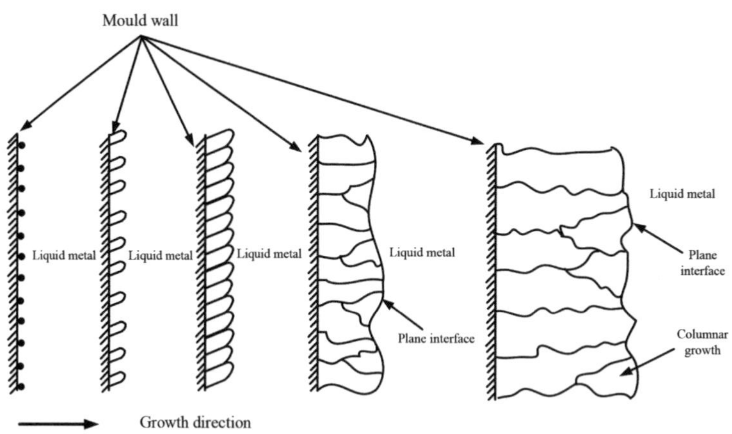

Figure 3.
Schematic presentation of planar growth giving rise to columnar dendrite.

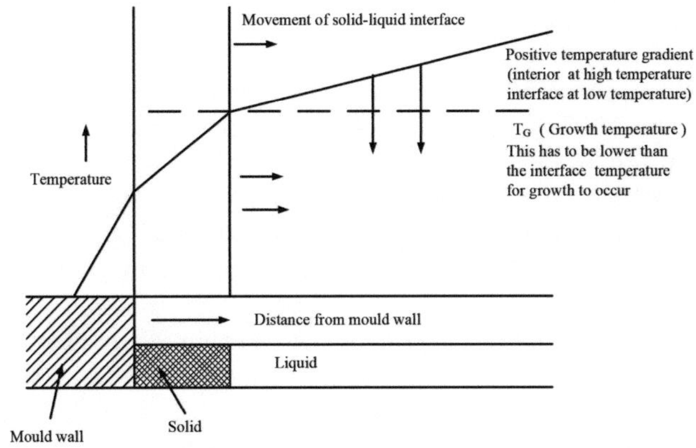

Figure 4.
Schematic presentation of the thermal condition for plane front growth.

temperature. This situation is favoured at slow cooling rates which ensure a stiff and positive temperature gradient. Under these conditions only, the interface assumes the shape of a seemingly plane surface a columnar grain-structure is favoured.

3.2 Thermal super-cooling

The thermal conditions get grossly distributed when sufficient accumulation of the liberated latent heat of crystallisation at the interface is experienced. The liberated heat now disturbs the thermal gradient. Though there is a positive thermal gradient due to the cold mould surface, the local evolution of latent heat produces a reverse temperature gradient at the interface. This is illustrated in **Figure 5**. Where a zone of thermal super cooling indicating pulls in the melt at the interface or adjacent to it at temperatures lower than the equilibrium temperature are witnessed. Obviously, growth does not occur due to the general advancement of the plane-front but by preferential growth processes in these undercooled pulls in the melt.

The planar growth pattern is disturbed as the minimum temperature in the liquid melt is not witnessed at the interface. Plane-front growth is hindered and growth occurs by other means. Depositions of further atoms on the surface of the nuclei may occur in regions of greater under cooling in preference to the interface. The thermal super cooling greatly influences the final structure of the solidified melt.

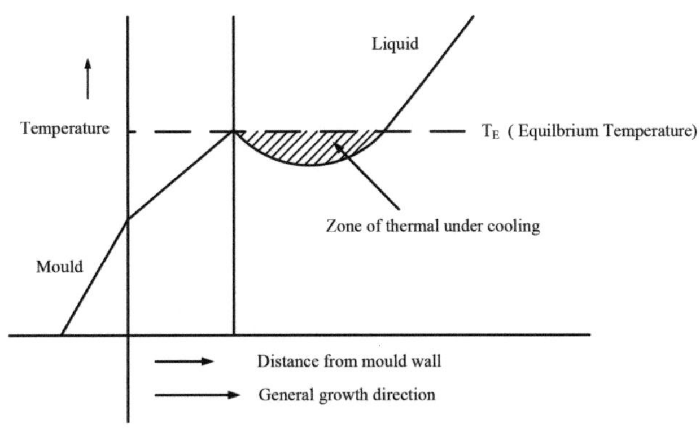

Figure 5.
Schematic presentation of generation of thermal undercooling.

3.3 Constitutional super cooling

Constitutional supercooling in an alloy is best illustrated in **Figure 6**.

The **Figure 6** presents the solidification and hence the phase changes in a simple binary alloy of 'A' and 'B'. Let us consider the alloy of C_0. The initial alloy deposited from Co has a composition confirming to 'C_1'. Obviously, 'C_1' has a composition pertaining to 'B' which is less than that of the original alloy 'Co'. Therefore, as 'C_1' is formed, the residual liquid gets slightly enriched in 'B'. Thus, as solidification proceeds 'B' is continuously rejected into the liquid. This rejection occurs at the solid–liquid interface throughout the process of freezing. A constitutional gradient is, thus, created in the liquid, solute 'B' being continuously rejected at the interface. The concentration of 'B' is maximum at the interface and gradually diminishes as one goes towards the interior of the liquid melt. This compositional variation is presented in **Figure 7(a)**. The change of composition brings in a corresponding change

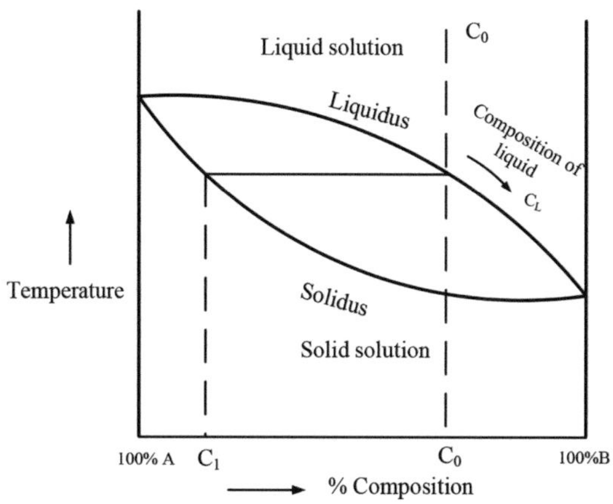

Figure 6.
Phase diagram of a typical binary alloy.

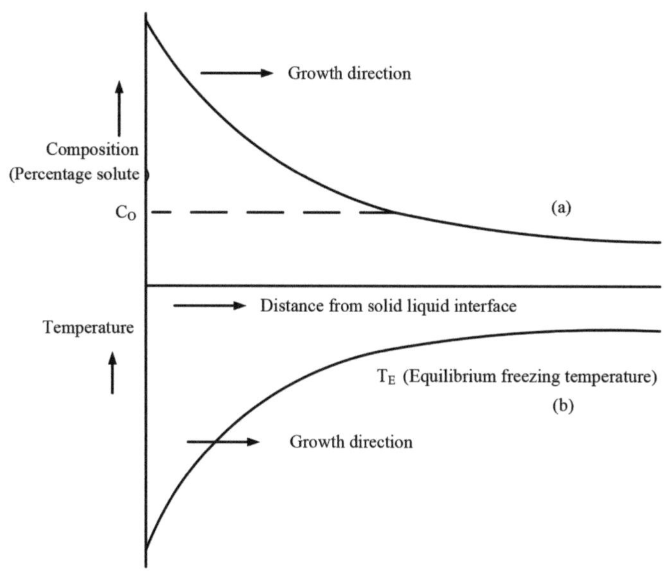

Figure 7.
Schematic presentation of rejection of solute at the interface (7(a)) and change of equilibrium temperature (7(b)) as a consequence of the solute accumulation.

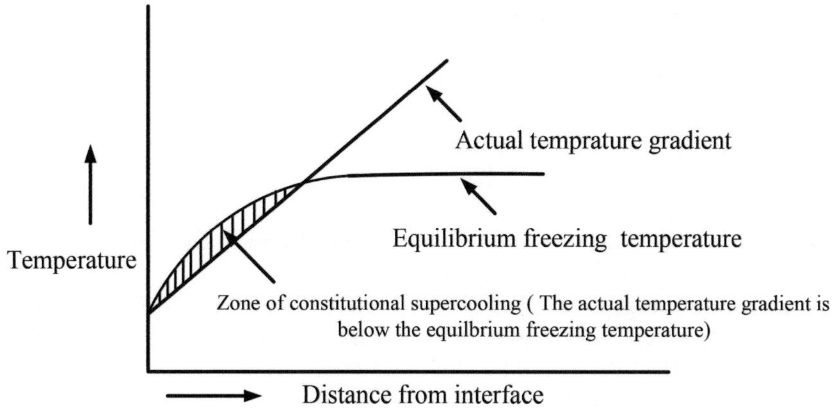

Figure 8.
Schematic presentation of constitutional supercooling as a consequence of solute rejection at the interface and the resultant alternation in the equilibrium freezing temperature.

in the equilibrium freezing temperature of the alloy as presented in **Figure 7(b)**. Each composition on the solute distribution curve has its corresponding equilibrium freezing temperature as it depends on the corresponding composition of an alloy.

The relationship between the actual (existing) temperature gradient in the melt and the equilibrium freezing temperature as a consequence of alterations in the composition of the alloy-melt, is illustrated in **Figure 8**.

Figure 8 clearly illustrates, before the actual (existing) temperature falls considerably for growth to occur, there, is a pool of melt where considerable supercooling can be witnessed at points farther within the melt. In this pool of supercooled liquid conditions are more favourable for freezing than at the interface. This condition is referred to as constitutional super-cooling.

3.4 Freezing of eutectic alloys

Alloys having Eutectic composition or containing appreciable amounts of eutectic constituents, undergo eutectic freezing. The alloy of eutectic composition solidifies at a single temperature to precipitate a mixture of two phases $'\alpha' and' \beta'$ of different compositions, 'x' and 'y' [2]. Thus, two separate phases, of two different compositions are precipitated out at a single temperature as a consequence of freezing of an eutectic alloy. Here, unlike the growth of grains as in a solid solutions, each of the eutectic grains is formed by the simultaneous growth of two dissimilar phases in close association. The eutectic structure could be any one of the following:

a. Alternate laminates of the two,

b. Rod-like or globular solids of apparently discontinuous phase in the matrix of the other phase.

3.5 Other growth modes

It is the extent of undercooling and its relative location in the melt that immensely influence the mode of growth of the crystal in a solidifying melt. As suggested earlier, it can be a thermal undercooling or a constitutional undercooling. Different extents of undercooling may be found in a band of liquid adjoining the interface or even in the inside of the melt depending on the following:

i. Temperature gradient in the melt,

ii. The equilibrium freezing temperature and

iii. The nucleation temperature (which will also be dictated by heterogeneous nucleation)

Depending on the extent of under cooling growth can be Dendritic Growth, Cellular Growth or growth due to independent nucleation.

Dentric crystalline growth takes place on solidification of a metal/alloy melt when the liquid–solid interface moves into supper cooled liquid at a temperature lower than that of the interface. This is illustrated in **Figures 5** and **6** wherein the thermal supercooling or the constitutional supercooling, as the case may be, generate pools in the liquid melt with temperature less than that at the interface. To understand dendritic growth it is important to realise that any protuberance on the solid face may tend to be stable and act as a centre for further growth in preference to other locations due to undercooling. The general advancement of the interface is retarded by the liberated lateral heat of crystallisation or by a solute barrier, but the local growth centres have the possibilities to grow into the zones of supercooling. This gives rise to dendritic growth. This is characterised by commercial alloys forming solid-solutions. It can be emphasised, under rapid solidification conditions non equilibrium condition of solid –liquid interface influence the dendritic characteristics to a great extent [12].

Primary axis of the dendrite is a result of preferred growth at the edge or corner of an existing crystallite. The projection develops into a needle, an then into a plate following the general direction of heat flow. This growth direction is usually associated with a particular crystallographic direction. Again, lateral growth of the primary crystal, needle or plate, is restricted by the liberation of latent heat of crystallisation or solute accumulations that had earlier restricted the growth of the original interface. However, the secondary or tertiary branches may grow by a similar mechanism that helped the formation of the primary stem. This is presented in **Figure 9** which depicts the branch like dendritic growth.

This unidirectional dendritic growth produces columnar dendritic structure.

In a pure metal dendritic growth is detected by interrupted freezing and decantation (once a portion freezes, it is separated from the liquid, i.e., the liquid is decanted from the freezing crystal). On the other hand, in alloys dendritic growth is revealed by the characteristic cored structure. Coring is resulted from the differential freezing processes. The centre of the dendrites are deficient in solute which are rejected to the interdendritic zone, as explained earlier.

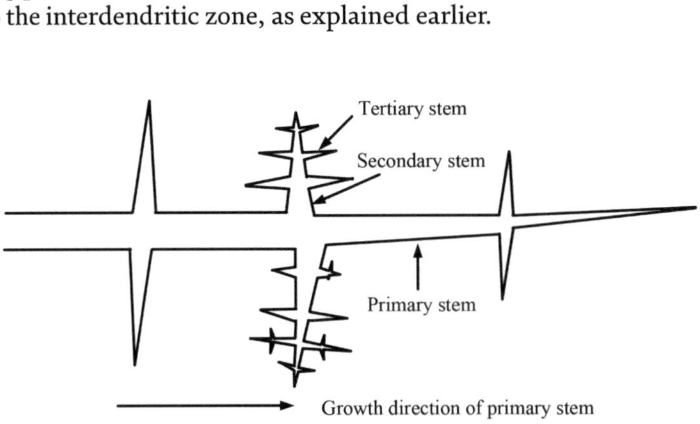

Figure 9.
Schematic diagram showing the growth of a typical dendritic arm.

Dendritic growth may be associated by crystals growing independently on independently formed nuclei, elsewhere in the melt depending on the preventing thermal conditions. This independently growing crystal within the melt has an interface on its periphery. Thus, it is capable of growing in all directions generating an approximately equiaxial grain. With a less marked undercooling, when the undercooling is not enough to form dendrites, cellular growth may still take place. Thus, cellular growth precedes dendritic growth. The cellular substructure is produced as a cluster of hexagonal rods. These rods grow into the liquid and reject solute on their boundaries at the respective interfaces. After a certain level of undercooling is achieved by both thermal and the constitutional means, cellular growth gives way to dendritic growth. This proceeds by the preferential development of some of the cells. This intermediate, rod like structure is also referred to as *Fibrous Dendrites*.

As shown in **Figure 10** when the temperature gradient is very shallow or the rate of freezing is very rapid, the undercooling achieved may be sufficient to promote nucleations at points in the melt,

distant from the main interface. In such an eventuality, the nuclei are free to grow in all directions on their periphery. An equiaxial grain structure, is thus, produced by independent nucleations. **Figure 10**, thus, exhibits the effect of increased undercooling (with the creation of different temperature gradient) on the mode of growth. It also shows the growth pattern with this different temperature gradients existing in the melt in the growth direction away from the mould wall into the interior of the liquid-melt.

3.6 The structure of the casting

Three factors have major influence on the casting structure.

3.6.1 Alloy constituents

The alloy constitution (composition) decides whether the structure will be of a simple phase or eutectic grains or both. The alloy composition also indicates the tendency of the alloy to respond to constitutional supercooling. The extent of constitutional supercooling is certain to influence the growth pattern that decides the crystallographic morphology of the casting.

3.6.2 Thermal conditions

The thermal conditions to which the liquid melt/alloy is exposed during solidification, refer to both, the rate of cooling and the temperature distribution in the

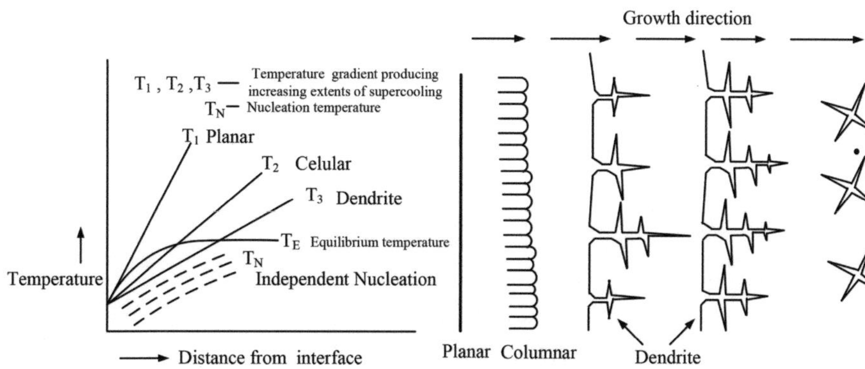

Figure 10.
Schematic representation of the influence of undercooling on the growth pattern and grain morphology.

solidifying melt. This is also related to the thermal properties of the melt as well as that of the mould. Obviously, the above would influence the cast structure by dictating the mode of growth.

3.6.3 Inherent nucleation and growth conditions in the liquid melt

The inherent nucleation and growth conditions in the melt are decided by the presence of foreign particles as well as the solute present in the melt. These solute atoms could be present as trace impurities or may be due to deliberate additions to influence nucleation. Obviously, these will influence/modify the possibilities of nucleation and growth, influencing the cast structure.

To elaborate the above we take help of **Figures 11** and **12**.

The alloys considered in these figures form a continuous range of solid solutions. The **Figures 11** and **12** illustrate the mode of crystallisation and hence the structure of the casting, as governed by the interaction of temperature and compositional gradients in the liquid.

Figures 11 and **12** depict the effect of temperature gradient and that of liquidus temperature profile, respectively on the structure of the casting. Initially when the melt is at higher temperature the existing temperature gradient is stiff [T_i (in **Figure 11**)] planar growth is encouraged and columnar grain structure is favoured. This is assisted by a slow cooling rate. This continues till the temperature gradient

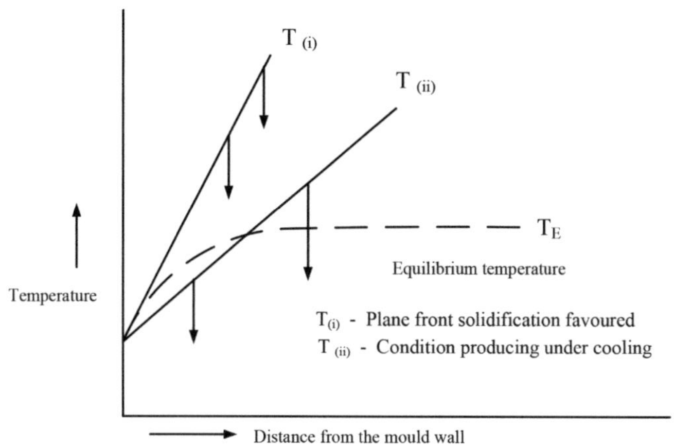

Figure 11.
Effect of temperature gradient variations on the extent of undercooling that influence the crystal structure.

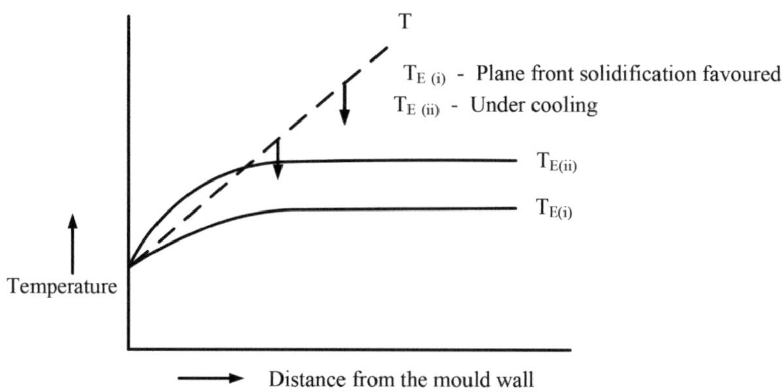

Figure 12.
Effect of liquidus temperature profile on extent of supercooling and crystal structure.

is sufficiently shallow to generate considerable undercooling which disturbs planar growth and growth proceeds adopting other modes, as explained earlier. **Figure 12** clearly indicates, with a given temperature gradient the alterations of equilibrium temperature profile, which could be due to the alterations in the solute concentration, undercooling is witnessed with liquidus profile TE (ii), with the liquidus profile TE (i) and the given temperature gradient 'T', undercooling is not witnessed and growth proceeds by plane-front growth giving rise to columnar grain structure. From the above, it can be concluded that columnar growth is promoted under stiff temperature gradients. Columnar growth is also favoured at slow cooling rates because of the following:

slow cooling rates establish low rate of nucleation in comparison to the growth rates, allowing growth to overtake nucleation.

As seen in **Figure 13** when the cooling rates are slow the solid rejected at the interface get sufficient time to migrate into the melt interior, away from the interface. The equilibrium temperature is altered.

It changes from TE (ii) to TE (i) (**Figure 13**). This is parallel to a situation as in **Figure 12** when with TE (i) the extent of undercooling are negligible or absent. Such a situation promotes columnar growth.

3.7 The temperature gradient (G) and rate of cooling (R) ratio (G/R)

In a foundry the various local factors like the extent of superheat, the extent of heterogeneous nucleation, the mould characteristics, etc. decide the variations in the thermal gradient (G) and the rate of cooling (R). Needless to say, the ratio G/R forms an important parameter to decide the mode of growth and the consequence of structure development.

Figure 14 illustrates that as the G/R ratio progressively changes from a high to a low volume the effect of undercooling becomes more and more pronounced. Columnar, plane-front growth gradually gives way to independent nucleation.

During freezing the thermal conditions prevailing in the melt continuously change. Thus separate structural zones as shown in **Figure 15** are encountered in the solidifying melt.

These zones are consequences of the continuously changing G/R ratio in the melt. Assuming of a lower value of the G/R ratio with the lapse of time results in the increasing extents of undercooling which are instrumental in the separable structural zones as preserved in **Figure 15**. The above can be made more clear with the aid of **Figures 16** and **17**.

Both **Figures 16** and **17** provide for an explanation of the mixed structure in a solidifying casting on the basis of the prevailing thermal conditions. To start with, as presented in **Figure 16**, the temperature gradient is stiff. Solidification initially occurs under this marked thermal gradient. This is often sufficient to cause

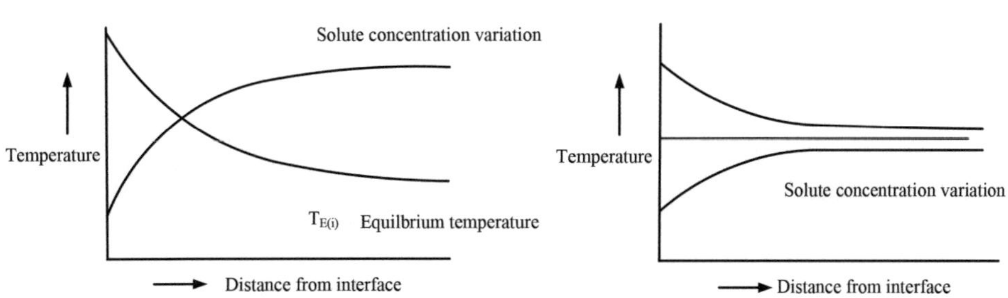

Figure 13.
Change in equilibrium temperature profile as a consequence of solute concentration crystal variation.

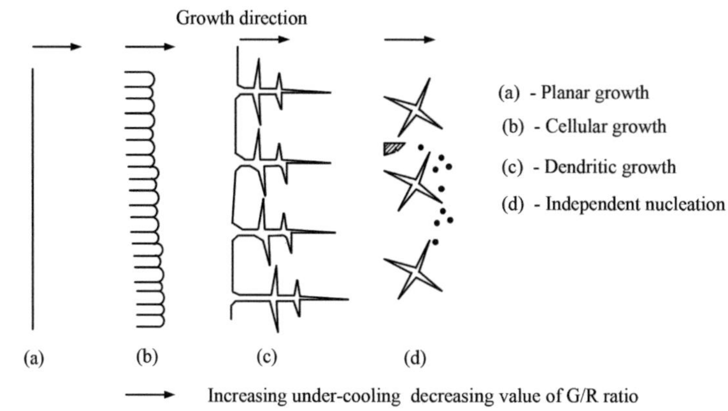

Figure 14.
Schematic presentation of G/R ratio influencing the effect of undercooling and the resultant structure.

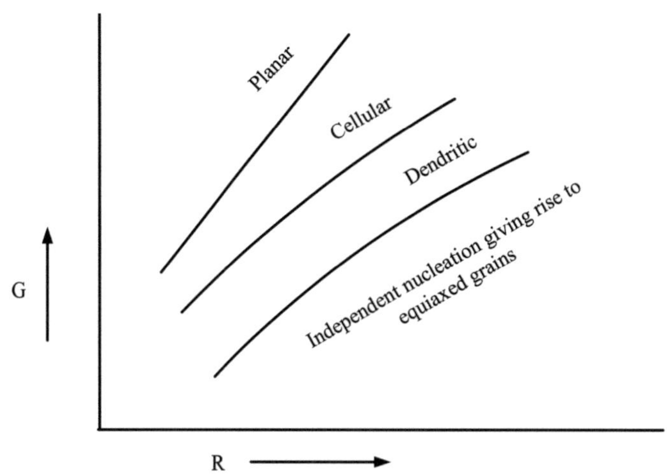

Figure 15.
Schematic presentation of critical changes in the G/R ratio during freezing and its influences on the different structural zones.

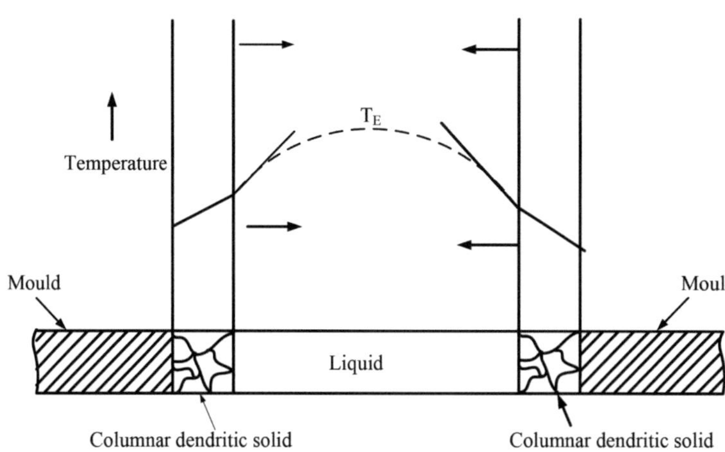

Figure 16.
Variation of undercooling with alteration in the thermal gradient showing different grain morphology.

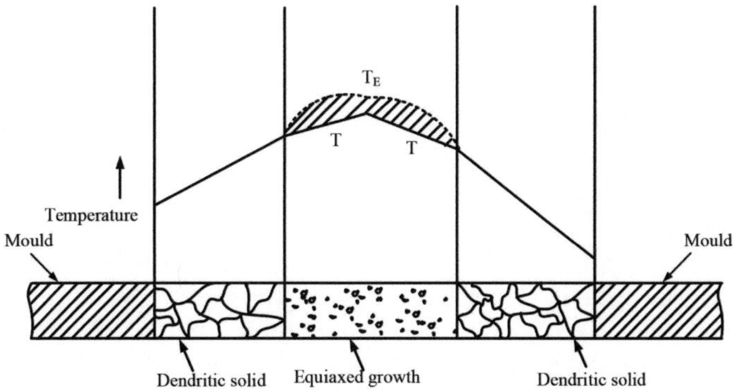

Figure 17.
Variation of undercooling with alternation in the thermal gradient showing different grain morphology.

columnar dendritic growth in the outermost region and adjacent to the mould wall as shown in **Figure 16** in this central zone (in some cases, throughout the entire solidifying melt) the temperature gradient is shallow (**Figure 17**).

This shallow gradient generates excessive undercooling. Here solidification proceeds by widespread nucleation, the rate of nucleation being very high. Independent nucleation occurs in the interior of the melt. These nuclei, without any barrier for growth across their periphery, grow into equiaxed grains. To be more specific, initially the temperature gradient is stiff. The rate of cooling is low, G/R assume high values. Initial solidification, thus, occur under a marked temperature gradient which is sufficient to cause columnar dendritic growth in the outermost layer. Gradually 'G' decreases, i.e., the temperature gradient becomes shallow and 'R' the rate of cooling increases as a consequence of increasing extents of undercooling. The shallow temperature gradient in the casting and the increasing extents of undercooling in the melt give rise to the formation of independent nucleation in the melt interior forming equiaxed grains, being free to grow on their unhindered periphery.

Schematically the practical, ideal cast structure can be presented as in **Figure 18**.

Although, the above refers to alloys forming solid solutions, analogous changes occur in alloys subject to eutectic freezing. The **Figure 18** shows small dendrites (equiaxed) in the outermost surface because of chilling effects at the cold mould wall.

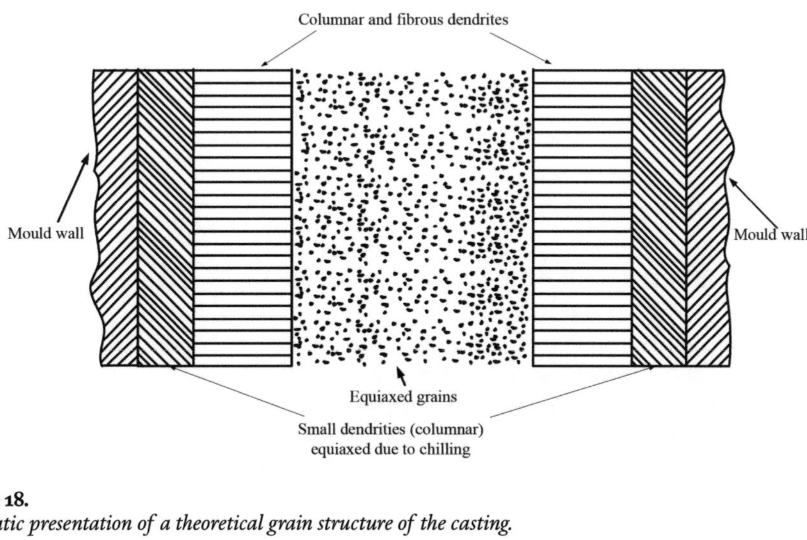

Figure 18.
Schematic presentation of a theoretical grain structure of the casting.

3.8 Further factors influencing the cast structure

Two further factors, other than the thermal and constitutional undercooling, also influence the cast structure particularly concerning the formation of equiaxed crystals. These are:

3.8.1 Crystal multiplication

Crystal multiplication is a consequence of the fragmentation of dendritic arms in the columnar zone due to local factors like thermal fluctuations and change in growth rates, etc. Also the nuclei formed on the inside of the mould wall may get washed off when further metal is being poured into the mould. Some of these detached nuclei may vanish being unstable, while some may get transported to favourable sides in the liquid and grow into equiaxed grains.

3.8.2 The transportation of these fragmented or detached crystal may occur by

i. Turbulence during pouring

ii. Thermal convective currents between the hot central portion and the relatively cooler surface region,

iii. Gravitational separation due to the difference in densities between the solid and the liquid. In general the fragmented dendrules in the melt tend to go to the bottom of the melt due to density differences between the solid dendrules and the surrounding liquid melt. The only exception is 'Be'. Here the density of the liquid is higher than that of the solid.

To sum up the formation of the equiaxed zone consisting of equiaxed grains, is promoted by the following:

a. Heterogeneous nucleation in situ

b. Crystal multiplication and

c. Transportation of crystallites by gravity or by mass movement into the interior of the melt

Obviously, the above form certain factors which influence the crystallographic morphology of the casting.

4. Mould feeding issues

For a casting to be produced, it is essential that appropriate technique to be adopted for the liquid metal to be fed into the mould cavity. It is an issue inviting special considerations since the viscosity of the melt increases with drop of temperature making its flow sluggish and time taking which may result in considerable solidification prior to the completion of the feeding process. Also, the metal/alloy shrink on solidification producing solidification shrinkages in the casting which are discontinuities in the casting. For a healthy casting production ample facilities must be made be made available for compensating for these shrinkage. On these considerations, we can take the case of a pure metal which solidifies at a constant temperature or even alloys having a narrow freezing range. In these cases, three are clearly defined interfaces between the solidified region and the 'still-liquid' region. Any solidification contraction has the liberty to be compensated by the 'still-liquid' melt adjacent to it. If sufficient liquid metal is available the process of compensation of the physical contraction continues by the general lowering of the free liquid surface resulting in the production of a sound casting with no solidification shrinkages. The supply of liquid metal is accomplished by the provision of liquid metal reservoir known as feeder head or riser.

However, in many cases the case is not as simple. A clearly defined solid/liquid interface does not exist. Solidification takes place through a zone simultaneously. Even in certain cases the solidification zone may extend throughout the melt entirely. In these zones crystals at different stages of growth can be seen with the residual low melting point liquid. The alloy is in a pesty zone or in a mushy stage. Contraction sites are dispersed in the casting making feeding of these contraction sites a very difficult and even impossible, for the production of sound casting.

Under these conditions, liquid metal is fed ink the contraction sites in three successive stages.

4.1 Stage I

In the early stages, growing crystal bodies are suspended in the liquid. Free movement of the liquid across the crystals is possible. Thus, any contraction can be easily compensated by a feeder head with the general lowering of the free liquid surface.

4.2 Stage II

Here, after some lapse of time, the grains grow to certain extent and form contact among themselves forming a network of solid. Liquid movement becomes confined to intergranular channels. These channels get diminished continuously. Frictional resistance to the liquid movement in these channels increases. It becomes considerably difficult for the liquid to reach the solidification sites that undergo contraction. Thus, feeding for compensating the contractions becomes progressively difficult.

4.3 Stage III

Now the final stages of solidification have reached. The intergranular channels are completely blocked by the growing crystals. Thus, isolated pockets of liquid are generated which solidify independently. The resultant contractions cannot be fed from external sources. These are always with long freezing range and cool under shallow temperature gradients. The shrinkage defects in these alloys persist resulting in spatter porosities distributed in the entire castings, even extending to the casting surface.

An external feeder head, known as the riser, is employed to compensate for the solidification shrinkages so that a sound casting results. For successful functioning of the riser the principle of directional solidification is employed. It is also known as progressive solidification in which solidification starts farthest from the riser and proceeds into the riser so that any side of shrinkage has an unfailing supply of liquid metal. The successful functioning of the riser must ensure the following:

i. Riser should be the last one to solidify in the casting system. This means throughout the process of solidification the riser must have liquid metal for feeding during freezing.

ii. Freezing must start farthest from riser and continue through the casting towards the riser.

iii. There must be a continuous path of feeding the liquid metal from the riser to the solidifying site.

The cooling rate of the casting need to be controlled for this purpose. The cooling rate on the other hand can be controlled by controlling the pouring temperature of the metal, pouring rate, promoting differential cooling by use of chills, differential heating by addition of exothermic materials, use of padding, etc. The temperature gradient and cooling rate are very important consideration in a solidifying melt. It is opined [13, 14] by setting up an appropriate temperature gradient and cooling rate by selecting the necessary pouring rate and temperature the cast structure can be controlled and the casting upgraded. A fine grain structure can be, thus, obtained by proper selection of the pouring rate and temperature. These fine grained structure can enhance the ability of the casting to inhibit the slide of the dislocations. This can result in the increase of yield strength and the ultimate strength of casting. These measures stiffen the temperature gradient assisting the setting up of a path feeding from the riser to the contraction sites in the mould.

5. Freezing characteristics of alloys

Though, no two alloys have identical feeding characteristics, on the basis of the major contrasts of solidification they can be put under three categories.

5.1 Gr.I

This group of alloys freeze with marked skin formation. These have of obvious short (narrow) freezing zone. These include Low carbon steels, Brasses, Aluminium Bronzes, Aluminium Copper, etc.

In these group of alloys progressive or directional solidification measures can be easily achieved. Sound castings can be obtained with proper feeding from the feeder head. However, the casting yield may suffer sometimes as the feeder head has to be finally discarded.

5.2 Gr.II

These are alloys with long feeding ranges. These include Medium and High carbon steels, Nickel based alloys, Gun metals, Mg alloys, Complex Al alloys, etc.

In these alloys solidification proceeds simultaneously in much of the casting, even in the entire casting. All the three stages of freezing can be clearly witnessed as explained earlier. In the third and last stage of freezing widespread porosity may occur. In such castings of alloys progressive solidification is very less unless heavy chilling is used to disturb the thermal gradient. The chilling induces very sharp temperature gradient and help formation of a sound casting by setting up of narrow freezing zones. However, the casting feeding in these alloys is not always based on directional solidification. Efforts are made to follow measures such that concentration of porosity is not localised but distributed in the casting. To achieve the above mentioned disperse porosity measures have to be adopted for equalisation of cooling rates over the entire zone instead of going for a sharp temperature gradient. These disperse micro pores could be more acceptable than concentrated porosities [15].

5.3 Gr.III

These are the alloys which show expansion on freezing. These include Grey Cast Iron. In the Gr III hypoeutectic Grey Cast Irons freezing is initiated with the growth of austenite dendrites. Contraction on freezing occurs much like the other alloys with considerable freezing range. Then the eutectic freezing begins. Graphite precipitates out of the solid. Interdendritic liquid gets enriched with carbon. The solidification of this austenite-graphite eutectic is accompanied by volume expansion. A positive pressure is caused. In a completely rigid mould this expansion makes it virtually self-feeding. In practice, however, the positive pressure tends to cause mould-wall-movement. This movement increases the mould dimensions and sets up a tendency for contraction giving rise to an increase in the internal porosity.

To sum up, in the Gr.I alloys feeding by risers is easy. In the Gr.2 alloys feeding by riser is helped by ensuring directionality in solidifications and in the Gr.III alloys, in the contraction stage all the three stages of bulk feeding and interdendritic feeding stages may be encountered.

6. Effect of temperature gradient on the feeding range of the alloys

As mentioned earlier the temperature gradient in a casting system can be made stiff from a shallow one by adopting several means. This is illustrated in **Figures 19** and **20** [16].

A stiff temperature gradient can reduce the extent of pasty zone with the associated advantages in setting of directionality in solidification. On the other hand, a shallow temperature gradient can be set up and extensive pasty zone resulting in simultaneous freezing in an extended zone in the melt and help distribution of micropores.

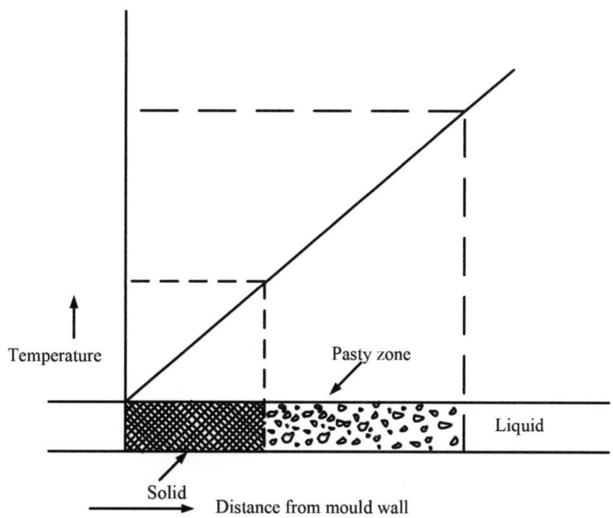

Figure 19.
Shallow temperature gradient showing extensive pasty zone.

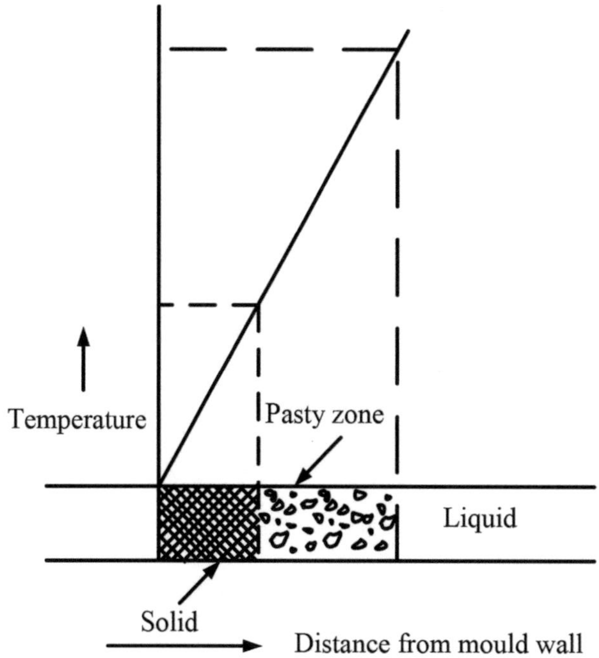

Figure 20.
Effect of stiff temperature gradient on pasty zone (narrow zone of crystallisation).

7. Concluding remarks

Casting of a metal/alloy is a manufacturing process in which the liquid metal/alloy Is poured into a pre-formed mould. The liquid solidifies in the mould and the solidified liquid, known as the casting, is finally retrieved from the mould. In the whole process of producing the casting, the solidification processes play a major role in deciding the cast structure which dictate the structure related properties of the casting and decide its end-use. Throughout the entire process of solidification the metal/alloy shrink, generating discontinuities in the casting in the form of shrinkage cavities. These have to the compensated for by the supply of liquid metal from a liquid-metal-reservoir, known as the feeder-head or the riser. For efficient functioning of the riser it is desirable to ensure directionality in solidification by shaping of the necessary thermal gradient 'G' and the rate of cooling 'R'. Infact the ratio 'G/R' plays a very important role in deciding the cast structure when it changes from a high value to a low value with the lapse of time. It decides the mode of growth which can be planar, cellular dendritic or growth due to independent nucleation and dictates the consequent development of the cast structure. The extent of superheat in the melt, the extent of heterogeneous nucleation, the mould characteristics, etc. Setup the required 'G/R' ratio. External factors like the pouring rate and pouring temperature can be altered suitably to vary the 'G/R' ratio such that the desire cast structure can be obtained.

DOI: http://dx.doi.org/10.5772/intechopen.94393

Author details

Upendra Kumar Mohanty and Hrushikesh Sarangi*
ITER, SOA Deemed to be University, Bhubaneswar, Odisha, India

*Address all correspondence to: hrushikeshsarangi@soa.ac.in

References

[1] Neil W Ashcroft; N David Mermin, Solid state physics, New York : 1976, Holt, Rinehart and Winston

[2] Smith, William F.; Hashemi, Javad, *Foundations of Materials Science and Engineering* (4th ed.), McGraw-Hill, 2006, ISBN 978-0-07-295358-9 p. 326-327

[3] Degarmo,E.Paul, Black,J-T, Kohser,Ronald.A, Materials and Processing in Manufacturing', (9th edition) Willy Publication, ,2003, p. 277

[4] Campbell,J., Metal Casting Processes, "Metallurgy,Techniques and Design", Complete Casting Handbook (Butterworth-Heinemann), 2015, 2nd edition

[5] Eskin,D.G. and Katgerman,L., Thermal Contraction During Solidification Of Alluminium Alloy, Materials Science Forum, 2006, Vols-519-521,p. 1681-1686,

[6] Oxtobody,D.W., 'Homogenous Nucleation: Theory and Experiment, Journal of Physics: Condensed Matter, 1992, Volume 4, issue 38, p. 7627

[7] Mahata,Avik;Asle Zaeem Mohsen and Baskes,I.Michael, 'Understanding Homogenous Nucleatiiion in Solidification Of Aluminum by Molecular Dynamics Simulations, Modelling And Simulation, In Materials Science snd Engineering, 2018, Vol.26, No.2, p. 1-31

[8] De Moor.P-Pe,Bleeten,T.A. and Van Sanken,R.A., In Situ Observation Of Nucleation And Crystal Growth In Zeolite Synthesis. A Small Angle X-Ray Scattering Investigation On Si-TPA-MFI, J.Phys.Chem B, 1999, Vol. 103(10), p. 1639-1650

[9] Binsbergen,F.L., 'Heterogeneous Nucleation Of Crystallisation', 'Progress In Solid State Chemistry', 1973, Vol.8, p. 189-238

[10] Heterogeneous Nucleation',By Cec-Science 2018.

[11] Kurz,Willfried; 'Plane Front Solidification' ;Book Chapter, ASM Handbooks, On Line, 2008, vol-15

[12] Trivedi, R. and Kurz, W; 'Dendritic Growth', International Materials Reviews, 1994, Vol. 39 (2), p. 49-74

[13] Flemings, M.C., 'Solidification Processing', McGraw-Hill, New York, NY, 1974, p. 34-44

[14] Church,N;Wieser,P;Wallace,J.K, Mod.Castings, 1966.49, 129

[15] 'Foundry Technology'P.R. Beeley,Butterworth snd Co(Publishers), 1979

[16] Mohapatra,S., Sarangi,H. and Mohanty,U.K, 'Effect Of Process Factors On The Characteristics Of Centrifugal Casting', Manufacturing Review, 2020, Vol- 7, Number- 26, p. 1-15

Chapter 3

Heat Transfer Studies on Solidification of Casting Process

L. Anna Gowsalya and Mahboob E. Afshan

Abstract

This chapter deals with the heat transfer characteristics between the cast and the mold. Generally the heat transfer behavior between the cast and the sand mold is used and all the three modes of heat transfer are studied. The heat transfer characteristics from the cast is at a faster rate for a die mold than for the sand mold. Since the sand mold is used for most of the industrial applications for the complex shapes of metal the heat transfer and the shrinkage behavior in solidification has to be understood perfectly. In this chapter, since the heat transfer mechanism and the shrinkage behavior of the metal in the sand mold is interrelated, hence were predominantly discussed.

Keywords: solidification, casting, heat transfer, shrinkage, IHTC

1. Introduction

The modern metal forming industry has taken complete advantage and benefit offered by the advanced techniques in order to remain in today's competitive market. The solidification modeling is a phase-change phenomena which is amazingly complicated as well as critical in many areas of science and engineering and also very vital in the field of automotive and aerospace applications. In the field of foundry engineering, when the molten metal is poured into the mold cavity, the metal solidifies and discharges heat into the mold, the metal shrinks due to which an air gap is formed in between the cast and the mold. This air gap acts as an obstruction for the heat flow from the cast to the mold and is to be found as one of the moving boundary conditions to be given as input for the casting simulation software. In the simulation of a solidification of the casting process, many parameters play a significant role responsible for the quality of the cast.

The data base for the properties of commonly used materials such as density, thermal conductivity, specific heat, solidus temperature, liquidus temperature, latent heat release etc., for the simulation of casting parameters need to be maintained by the industries.

2. Heat transfer mechanism in solidification

To comprehend the heat transfer mechanism we need to know the behavior of solidification. The heat transfer from the liquid hot temperature cast to the mold is a very complex phenomenon and different modes of heat transfer can be observed while solidification in the cast. While heat transfer is predominant the resistance to the heat flow also has different dimensions to this solidification. This resistance mainly depends on liquid cast metal, latent heat release, interface, solidified cast, the type of mold and the ambient conditions. General solidification of an alloy is discussed in the **Figure 1** and specific cooling curve for Al6061 is shown in **Figure 2**.

Initially on pouring the liquid metal cast into the mold cavity the whole metal fluid flows and occupies the mold cavity, the liquid metal flowing with the velocity, mixes thoroughly and releases heat to the mold due to the very high temperature difference. Complete thermal contact is observed between the cast and the mold which causes the heat transfer to be purely conduction, where the resistance offered by this liquid metal is negligible since the entire fluid flow is the superheated cast metal. Once the cast metal reaches the liquidus point on cooling, the cast shrinks and releases latent heat and also a number of metal oxides are released which causes an air gap between the cast and the mold. Due to this air gap the heat transfer phenomenon now changes to a complex one where all modes of heat transfer can be observed simultaneously. This air gap is characterized with an Interfacial Heat transfer Coefficient (IHTC) "h" across the metal-mold interface. The rate of heat at the interface is found using the surface heat flux as q (W/m^2) and given by the Eq. 1.

$$q = h(Tc - Tm) \tag{1}$$

Tc and Tm are the cast and mold surface temperatures at the interface in K or deg. C.

The dynamics of solidification of cast metal, mold temperature and the cast temperature can be clearly understood from the cooling curves shown in **Figure 2**. Once the molten metal fills the cavity the alloy cast reaches the maximum temperature. Generally the heat transfer analysis starts from this point onwards as the temperature drops from the liquid cast metal to the liquidus temperature (T_L), the point at which the solidification begins and this freezing is called liquid cooling. The loss of superheat temperature of the cast metal after pouring is found due to

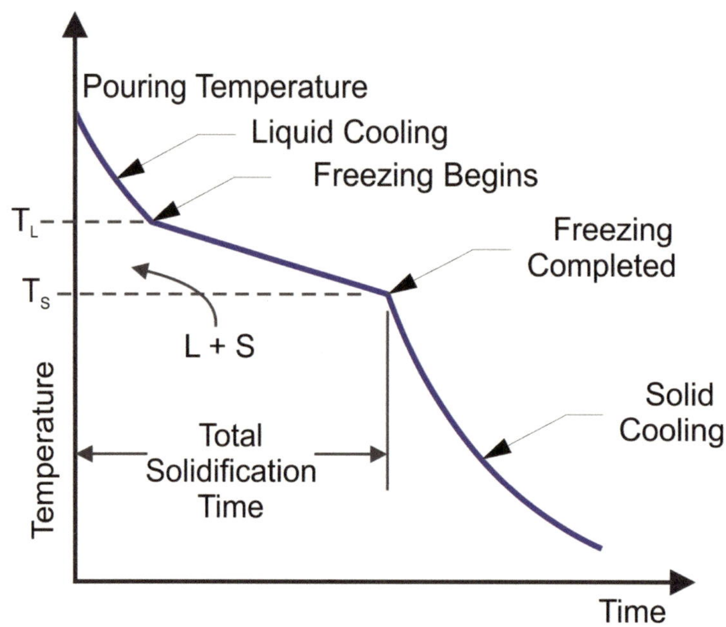

Figure 1.
Solidification curve for alloy.

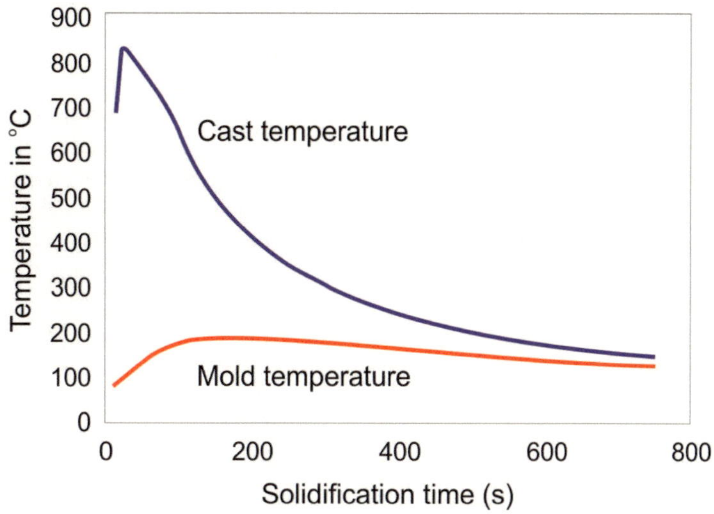

Figure 2.
Aluminum alloy (Al6061) solidification curve.

the turbulence in the liquid metal. This rate of cooling is linear and a minimum amount of heat is transferred from the cast to the mold as it is having a complete contact with the mold surface.

As the solidification progresses with time it reaches the liquidus point at the same time where the mold temperature increases significantly to a maximum temperature. Further the solid skin forms on the outer cast surface, the metal shrinks and an air gap starts forming between the metal and the mold. When the cast solidifies further the air gap separates the two surfaces. This is a common phenomenon in most of the alloys. The rate of heat transfer from the cast to the mold is very high as it releases larger quantity of latent heat to the mold and the cast temperature gradually reaches a solidus (T_S) temperature of the alloy. The air gap plays a significant role in varying IHTC with various factors influencing solidification.

Further solidification reduces the cast surface temperature, however the inner cast metal shrinks and it further releases the heat to the mold and there is rise in the mold temperature as shown in **Figure 2**. Thereafter further reduction in the cast temperature after the solidus point (Ts) was found as the third stage of solidification. The air gap size is further increased as the solidification time increase and its effects are felt till the end of solidification. However there is still a temperature difference between the cast and mold for the further heat transfer to continue.

Once the complete air gap is formed between the cast and the mold, the gap will contain almost all kinds of gaseous except air that contradicts the air gap term. The sand mold which is used for the casting application, generates the mold gases which are often high in hydrogen, containing typically 50 percent which fills the air gap. The hydrogen gas thermal conductivity increases the heat transfer by 7 times more as the mold temperature rises to a high temperature of 500°C due to radiation. Therefore it is very essential to know or analyze the interface during the solidification process as it is further discussed in the next section.

On comparing the green sand mold with dry sand mold the green sand mold expand homogeneously and release heat to the surrounding which leads to a lesser resistance for the heat flow whereas dry sand mold offers more resistance than the green sand mold. The high thermal conductivity die mold material has uniform temperature variation and assumes homogeneous expansion.

3. Shrinkage behavior of casting

While melting the metal in the furnace has a higher specific volume hence it occupies more space by the metal and on pouring it results in the solidification in the mold which increases the complexity of the solidification [1]. After pouring the temperature of the cast reduces and the specific volume also reduces which causes shrinkage in the poured volume as shown in **Figure 3**. To understand the complex behavior of solidification we need to understand three different stages of shrinkage of metal during the solidification process; it includes liquid shrinkage, liquid- solid shrinkage and solid shrinkage.

3.1 Liquid shrinkage

The superheated metal which is poured in the liquid state has more specific volume than the liquid metal in the cavity [2]. This liquid metal occupies the mold cavity and is in superheated state and comes in complete contact with the mold surface. Here the mode of heat transfer is purely conduction shown in **Figure 3**. On solidification there is a liquid contraction due to reduction in specific volume, the metal cools further and reaches to a liquidus temperature. This contraction of liquid metal separates cast and mold surface and imitates the air gap formation which is assigned as liquid shrinkage.

3.2 Liquid- solid shrinkage

Actually the liquid contraction leads to a solidification which is a complex problem in the casting industry. This requires a proper feeding mechanism to fill the cavity by maintaining high liquid cast temperature while pouring and if not then the partial liquid - solid contraction leads to shrinkage porosity. The specific volume of the solid metal is lesser than the liquid metal. All the solidifications are planned for the directional solidification which refers to the faster cooling rate at which solidification progresses from the cavity metal to the feeder mechanism. The faster cooling

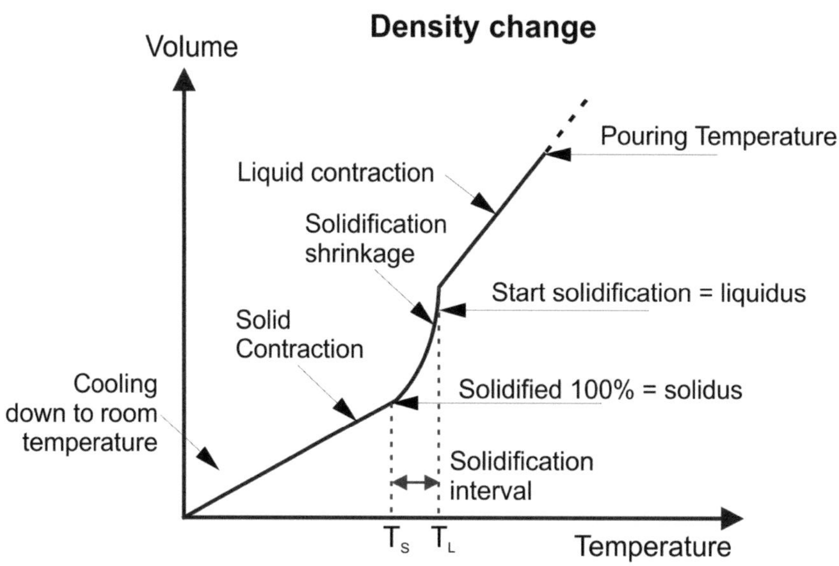

Figure 3.
Specific volume changes against cast surface temperature.

rate and the movement of liquid in the solidification is due to the area of the surface which enables the liquid metal to drop its high temperature to solidus temperature. The runner, riser and the gating system is designed in the mold pattern enhances the directional solidification by transferring proper heat flow from the cast to the mold.

The alloys of eutectic type allow lesser solidification shrinkage volume and also have a lower sensitivity to the solidification problems caused by sudden geometry changes. While they involve smaller risers, these can be omitted completely in certain cases by gates placed strategically and because the metal feed avenues stay open longer, it ensures a uniform solidifying process. While eutectic type of solidification is the most simplest, it requires the least reciprocity and can withstand a range of geometries. Directional solidification is more complex; however, when it has an ideally designed geometry, it is highly capable of extremely higher interior unity. Heat transfer is in fact the main process behind the bilaterally symmetrical and mutual state of connectedness in the process of solidification shrinkage and geometrical patterns. The heat transfer during solidification of castings involves three modes of heat transfer, namely radiation, conduction and convection, the rate of heat transfer is still dependent on the geometry of the casting as discussed later in the interfacial heat transfer coefficient section.

3.3 Solid shrinkage

The final stages of shrinkage in the solid state which can cause a separate series of problems. As cooling progresses, and the casting attempts to reduce its size in consequence, it is rarely free to contract as it wishes. This stage of solidification is usually complex either by the types of mold, or by the other casting parts like the runner and riser that have already solidified and cooled as the air gap formed. The air gap formed is mainly due to the various factors like metal oxide formation, coefficient of thermal expansion, latent heat released, evaporation moisture in the case of sand mold, interfacial gap, mode of heat transfer etc. this type of solidification shrinkage is also called as pattern shrinkage.

These factors are the major causes for the heat to flow from the cast to the mold and it is found that it majorly affects the solidification and in turn affects the quality of the cast product. The amount of solid metal stretches like plastic casting, makes the solidification again into a complex problem. This shrinkage behavior leads to difficulty in predicting the size of the pattern since the degree to which the pattern is made oversize (the 'contraction allowance' or 'patternmaker's allowance') is not easy to quantify. This shrinkage also causes hot tearing or cracking of the casting which lead to more localized problems.

In general, liquids contract on freezing because of the rearrangement of atoms from a rather open 'random close-packed' arrangement to a regular crystalline array of significantly denser packing. The densest solids are those that have cubic close packed (face-centred-cubic, fcc, and hexagonal close-packed, hcp) symmetry. Thus the greatest values for contraction on solidification are seen for these metals.

4. Interfacial heat transfer coefficient (IHTC)

The heat transfer characteristics during casting are governed by IHTC. The molten metal is poured into the cavity it first enters the mold due to the fluidity of the metal, it occupies the cavity and ensures complete contact between the metal and the mold. In the early stage of solidification, the fluidity of the molten metal conformance and contact between the cast and mold surfaces is good. At this early stage of solidification due to the nucleation of the metal, higher initial surface heat flux is reached. Further the solid skin forms and then spreads to cover the entire casting surface. As the solidified layer forms with sufficient strength, simultaneously air gap forms and as a consequence the contact between the casting and the mold are reduced. This leads to the sudden drop in the heat flux and the solid skin forms on the outer cast surface [3]. The cast liquid - solid shrinks/contracts away from the mold surface. This further releases heat and it is absorbed by the mold surface and in turn increases the temperature of the mold as it expands. The mode of heat transfer is not only due to conduction at this stage because the heat from the metal to the mold takes place across the interface region but also due to other modes of heat transfer convection and radiation. The air gap varies for the different cast metals and depends on their factors of the release of metal oxides, hydrogen gases and material properties of the cast and mold, geometry etc.

Further the third stage of solidification is identified between the liquidus to solidus temperature of the cast as the fall in the casting surface temperature is suddenly halted, due to the release of latent heat. After the complete solid skin formation on the cast the heat transfer further diminishes and gap size increases and the mode for heat transfer is assumed to be conduction of heat through the gaseous phase in the interface using the air gap method. This air gap size is measured as x by assuming the expansion to be homogeneous, and the interfacial heat transfer coefficient is estimated as $h = k/x$: where k is thermal conductivity of the air (W/mK) as shown in **Figure 4**. This concept of conduction as a mode of heat transfer in IHTC is reported by Kai- Ho and Robert D Pelhke, [4]. There are many factors that influence the IHTC and practically the IHTC becomes highly unpredictable if all the factors are not taken into account while designing. The various factors listed by the authors Lewis and Ransing, [5] and Guo Zhi-Peng et al. [6], that affect the interfacial heat during solidification is listed below.

1. Die coating thickness: The initial high peak value of IHTC is reduced with an increase of die coating thickness. While pouring the metal at the liquid stage the effect of die coating behaves as a weaker influence at the interface as the air gap formed.

2. Insulating pads, chills, etc.: The IHTC has different behaviors with insulating pads and chills. It is obvious that always the insulating material reduces the IHTC and the chills increases the IHTC.

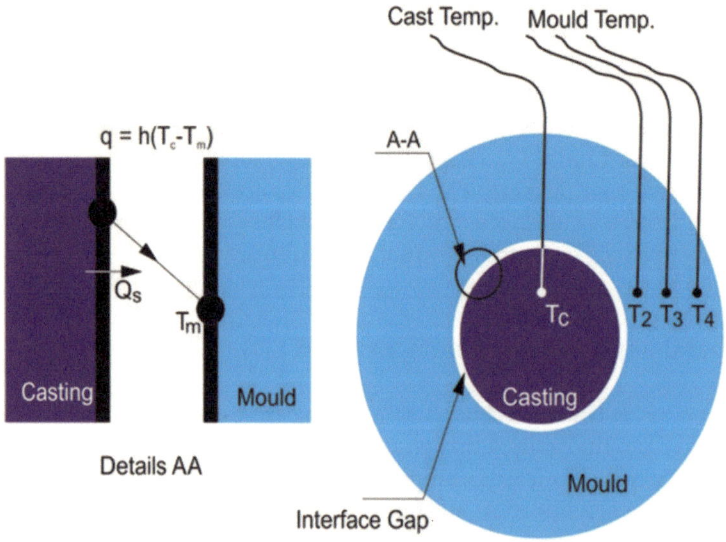

Figure 4.
Schematic representation of IHTC during solidification of casting.

3. Geometry of Casting: The area of contact with the mold and the directional solidification will have higher IHTC.

4. Pouring temperature: Higher values of superheat will increase the initial value of IHTC.

5. Surface roughness: Higher initial value of IHTC for the better contact when the surfaces are smooth.

6. Alloy composition: Higher initial value established for an alloy with a larger freezing range.

7. Latent heat: Cast from superheat temperature to liquidus temperature ensures sharp slope in IHTC due to the evolution of latent heat.

8. Metallostatic pressure: During the pouring of molten metal into the cavity rises the metallostatic pressure, this is also responsible for higher IHTC at the initial stage.

9. Mold temperature: During initial stage higher IHTC due to the higher mold temperature and smaller temperature difference for higher peak heat flux.

10. Die Coating thickness: Increase of die coating thickness decreases the IHTC. While pouring the metal at the liquid stage the effect of die coating behaves as a weaker influence at the interface as the air gap formed.

11. Mold materials

12. Type of castings

As it is pointed out by many researchers the gap size mainly depends on the gas that is formed in the interface. The rate of solidification of castings made in a sand mold is generally controlled by the rate at which heat can be absorbed by the mold. In fact, compared to many other casting processes, the sand mold acts as an excellent insulator, keeping the casting warm. However, of course, ceramic investment and plaster molds are even more insulating, avoiding premature cooling of the metal, and aiding fluidity to give the excellent ability to fill thin sections for which these casting processes are renowned. It is regrettable that the extremely slow cooling can contribute to rather poorer mechanical properties.

Extensive literature reviews have been made, in order to determine the interfacial heat transfer behavior during the solidification of casting at the metal-mold interfaces, since the 1970's. The boundary conditions as a surface heat flux and mold surface temperature established at the metal mold interface were used to determine the precise interfacial heat transfer coefficient value by using many mathematical methods described in the literature. The most common approaches can be distinguished here as follows for the determination of IHTC at the metal-mold interface including surface heat flux and mold surface temperature:

1. Air gap measurement technique

2. Pure Analytical approach

3. Semi-analytical method

4. Numerical Methods

The following section explains the detailed procedure of these methods listed above.

4.1 Air gap measurement technique

This method calculates the IHTC based on entrapped gas properties present at the interface. The thermal conductivity of the air between the cast mold interface and the distance of air gap measured as x with the LVDT [7]. The formula used for IHTC calculation is, $h = k/x$, W/m^2K. The mode of heat transfer assumed in this method is conduction at the interface, but the other modes of heat transfer are also practically possible as we have discussed in the above section. Hence this method is not widely accepted by the researchers.

4.2 Numerical approaches (inverse method)

In this approach, experimental cooling curves were obtained at certain locations of the cast surface and on the mold to estimate the IHTC. The IHTC is calculated based on measured cast temperature, estimated mold surface temperature and estimated mold surface heat flux. Generally solidification heat transfer problems as shown in **Figure 5** were categorized as

- Direct Heat Conduction Problem (DHCP)

- Indirect Heat Conduction Problem (IHCP)

In the DHCP the boundary conditions were known at the metal mold interface (which is a moving boundary problem and is difficult to acquire the parameters at the interface) and the effects were determined, mathematically it is known as a well posed problem. But in solidification of casting, knowing the boundary condition is very difficult because of its high transient nature, moving boundary problem, high temperature region, combination of all modes of heat transfer, etc., at the interface. So the inverse heat conduction problem is used to approach the problem. In order to calculate the boundary condition at the interface as a surface heat flux and surface temperature of the mold, experiments were carried out to determine temperatures in the mold to get the input data. This leads to a method of adoption of an ill-posed problem or the inverse heat conduction problem (IHCP) [8]. This ill-posed nature makes IHCP conduct experimentation to determine the boundary conditions at the interface before it has to be solved from the available data rather than using a DHCP approach.

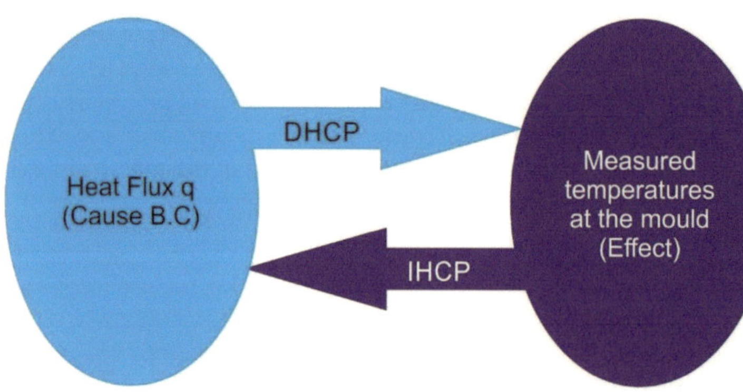

Figure 5.
Schematic diagram for DHCP and IHCP conditions.

The interfacial heat transfer coefficient at the cast mold interface can be calculated based on Eq. (1), requiring the transient surface heat flux. Cast and mold surface temperatures are measured using thermocouples during solidification regardless of its uncertainty in the physical measurements. The pure analytical or other methods mentioned above are unable to determine the surface heat flux at the interface. This leads to the numerical approaches and their formulation of inverse heat conduction problem (IHCP) at the interface to determine the boundary conditions. The boundary conditions at the interface are explored or determined by the IHCP. This has been studied by various techniques like FDM, FEM, FVM and CV methods. One of the common and mostly used method is mainly based on the function minimization technique based on the numerically calculated and measured data [6].

$$F(h)=\sum_{i-1}^{N}(T_i - Y_i)^2 \tag{2}$$

Where, F(h) is the minimization function, Ti, Yi are calculated and measured transient temperatures at the same locations, i= 0 to N, nodal point. The errors in the temperature measurement may also lead the IHCP into ill-posed. This problem leads the researchers to propose many techniques to solve for IHCP to determine boundary conditions at the interface with the measured temperature histories.

1. Polynomial extrapolation method: The temperature at the interface was deduced by extrapolating any one of the polynomial curve fitting techniques. This method needed many measurements inside the cast and mold surfaces. This mathematical tool failed to minimize measurement errors.

2. Regularization method: In order to minimize the error from the measurement obtained a sensitivity analysis can be carried out using the Tikhonov regularization theory. This was used to regularize some function to relate the measured data and this was improving the accuracy and stability of the results obtained. This method could achieve an excellent solution and could be applied to any complex geometry, but the computation takes a very long time.

3. Boundary element method and Laplace transform: the unknown temperature were transformed into equations as well as written as matrix format. This could be easily solved and written into a computer program. But it has some restrictions. It was an effective method to solve a simple linear problem. But the measured temperature data always has more noise (disturbances) in the data, this could fluctuate the result obtained as heat flux.

4. Beck's function specification with finite difference method (implicit & explicit): It was another minimizing error technique used based on heat flux, where sum of squares of assumed and calculated data are used into the function. This method could be used for linear or nonlinear problems. Also, it has long computation time and also could achieve an accurate solution with efficient computation.

5. Control volume method: This method works, based on energy balance applied over a control volume drawn on each nodal point. The next one is the governing equation for the transient heat conduction written as a partial transient heat conduction equation changed into an ordinary differential transient equation. This involves both energy and mass conservation on each node, leads to a complex formulation equation containing up to 4th order, which may be difficult to program using computer languages, and can only be applied to simple geometrical shapes and one dimension.

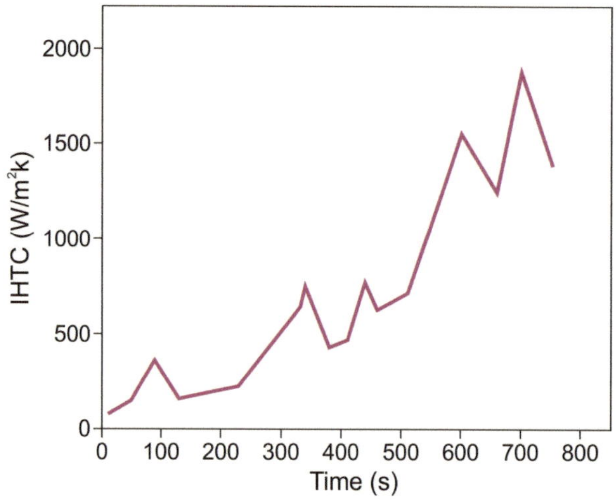

Figure 6.
IHTC variation for the rectangular aluminum casting with sand mold.

A sample of a rectangular geometry with an aluminum (Al6061) cast volume of 45 cm^3 was solidified and the IHTC was calculated as shown below in **Figure 6**. Here the IHTC curve was calculated using the control volume method and it shows a gradual increase. Various characteristics of the IHTC and the heat transfer can be discussed [9].

4.3 Behavior of IHTC for the given Al 6061

The behavior of the sample rectangular cast was considered as it summarizes most of the heat transfer modes in solidification of the cast. On pouring the IHTC was found to be 370 W/m^2 K at 90 s, the higher initial surface heat flux was due to a perfect thermal contact. As further solidification starts, vaporization takes place in the sand mold because of the moisture content, presence of hydrogen release along with metal oxides across the interface and the reduction of specific volume of metal creates an air gap and decreases the value of IHTC rapidly to a minimum value of 163 W/m^2 K at 130 s. The shrinkage of metal causes release of latent heat and rise in the IHTC, then heat transfer reduces once the solid skin is formed [10]. Again the inner metal leaks and flows out from the solid skin to outside and gets cooled which again releases latent heat and so IHTC increases and decreases. Continuous rise and fall of the IHTC shows peak formation, which is shown till the end of solidification. The fourth peak value of 1718 W/m^2 K at 600 s and further again at 720 s the IHTC reached the highest peak value of 1918 W/m^2 K. The vapor pressure developed in the sand mold is due to the escape of moisture content to the ambient, which is sufficient to allow the heat to flow from the solidifying metal to sand mold hence the sharp rise in IHTC is observed in the final stage of solidification. Not only vapor pressure but also huge temperature differences causes high heat flows. Due to the thermal resistance induced, as the metal solidifies and contracts, a fall in the IHTC is vividly observed.

5. Conclusion

The materials that change phase during solidification to room temperature can be much more complicated. The heat transfer in the solidification is a complicated phenomenon as shown in the above sections. Understanding the heat transfer characteristics while solidification will help to link the various developments in the micro structure of the materials and the dislocations present. When solidification is complete the strength of the material can be assessed and the formation of the grains in the material can be directed by control of the temperature and heat flow on solidification.

The IHTC of a sample of Al6061 is thoroughly explained to comprehend the various modes of heat transfer while solidification is taking place. Proper cooling helps to govern the solidification and as the temperature is sufficiently low the strains of dislocations will not be sufficiently mobile to migrate into low energy positions, forming low-angle boundaries. Thus the alloy will become sufficiently strong to retain any further strain as elastic strain. Once the metal solidifies properly the structure of the alloy will no longer be affected during further cooling. Hence a complete idea of IHTC at all the times of solidification is the best option to minimize the errors and maximize the strength.

Author details

L. Anna Gowsalya[1*] and Mahboob E. Afshan[2]

1 Department of Production Technology, Madras Institute of Technology, Anna University, Chromepet, Chennai, Tamil Nadu, India

2 Department of Mechanical Engineering, B.S. Abdur Rahman Institute of Science and Technology, Vandalur, Chennai, Tamil Nadu, India

*Address all correspondence to: gowsalyamahendran@gmail.com

References

[1] John Campbell: Castings, Butterworth Heinemann, Oxford, 2002.

[2] Bhagavath S, Cai B, Atwood R, Li M, Ghaffari B, Lee P.D, Karagadde S: Combined Deformation and Solidification-Driven Porosity Formation in Aluminum Alloys, Metallurgical and Materials Transactions A, 2019; 50, 4891-4899.

[3] Rajaraman R, Velraj R: Comparison of interfacial heat transfer coefficient estimated by two different techniques during solidification of cylindrical aluminum alloy casting, Heat Mass Transfer, 2008; Vol. 44, 1025-1034.

[4] Ho F, Pelhke R.D: Metal –mould interfacial heat transfer, Metallurgical Transactions B, 1985; 16, 585-594.

[5] Lewis R.W, Ransing R.S: A Correlation to Describe Interfacial Heat Transfer during Solidification Simulation and Its Use in the Optimal Feeding Design of Castings, Metallurgical and Materials Transactions, 1998; 29B, 448.

[6] Guo Z-P, Xiong S-M, Murakami M, Matsumoto Y, Ikeda S: Study on interfacial heat transfer coefficient at metal/die interface during high pressure die casting process of AZ91D alloy, China Foundry, 2007; 4, No.1.

[7] Griffiths W.D: The Heat-Transfer Coefficient during the Unidirectional Solidification of an Al-Si Alloy Casting, Metallurgical and Materials Transactions B, 1999; 30 B.

[8] Anna Gowsalya L, Jeyakumar P.D, Rajaraman R and Velraj R: Estimation and validation of interfacial heat transfer coefficient during solidification of spherical shaped Aluminum alloy (Al 6061) casting using inverse control volume technique, Frontiers in Heat and Mass Transfer (FHMT), 2019; 12, 21.

[9] Rajaraman R, Anna Gowsalya L, Velraj R: Estimations of interfacial heat transfer coefficient during the solidification of rectangular aluminum alloy casting using two inverse methods, Frontiers in Heat and Mass Transfer (FHMT), 2018; 11, 23.

[10] Zhang A, Liang S, Guo Z, Xiong S: Determination of the interfacial heat transfer coefficient at the metal-sand mold interface in low pressure sand casting, Experimental Thermal and Fluid Science, 2017; 88, 472-482.

Chapter 4

Simulation and Validation of Castings in Shop Floor

Partha Haldar and Goutam Sutradhar

Abstract

Production of sound casting demands a thorough understanding of whole casting process. But still, defects and rejection of castings are ubiquitous because in general, the designer lacks domain knowledge about casting processes and hardly have any methodology to find out the parameters that produce sound casting. Casting simulation software simulates the way how casting engineers decide the casting process in a virtual platform and also analyzes each decision to point out the design modifications needed to enhance the quality of casting as well as reduce lead time, tooling and manufacturing costs. The application of simulation software enables us to say, "Get it right, the first time and every time". Simulation software can be very helpful in calculating tedious formulas, constructing solid modeling which will be helpful to visualise the actual situation like core/mould assembly, gating and feeding arrangements with the main casting before going into actual practice. It can be adopted for troubleshooting existing castings, and for producing new castings without or minimum shop-floor trials. This chapter illustrates the advantages of casting simulation (both tangible and intangible), bottlenecks (technical and resource-related), and some best practices to subdue the bottlenecks. In this chapter some of the live examples have been cited to understand the process logically and scientifically.

Keywords: casting simulation, concurrent engineering, design for manufacture, solid modeling, quality assurance

1. Introduction

Simulation imitates a real phenomenon by the use of certain mathematical equations. Metal casting is a manufacturing process where molten metal is poured into a mould cavity of required shape and size and allowed to solidify. Naturally, metal casting simulation is a very complex phenomenon which involves flow of fluid, heat transfer between mould and molten metal etc. It is often said that the development of accurate simulation software is a 'rocket science for rocket scientists'. Actually, metal casting is a process which has numerous associated controlling factors. Therefore, the key to develop a practical useful casting simulation software is to figure out the related most important parameters. Several researchers have worked hard for several decades to find out the same. Geometry, material, and process are three major influencing factors related to metal casting [1].

The casting simulation software producing farms always keep target to accurately simulate the physical phenomena as far as possible like the mould filling, associated heat-transfer, solidification pattern of the metal/alloy, and the involved phase transformation of castings [2]. It is ubiquitous that a set of governing equations are required to model these phenomenon in a computer program. Now a days, these modeling methodologies are so strong that they can accurately predict the microstructure and mechanical properties of the castings. It can also pin point the position of internal defects like shrinkage porosity, sand inclusions and cold shuts etc. The simulation software can be used for the development of any new castings or it can be used for standardize any existing casting for any design change to improve yield of the casting without shop floor trial. For these reasons, casting simulation has become an indispensable tool in modern foundries. No foundry can produce high-quality castings particularly integrated castings without simulation for the first time. The present dynamic market demands fast response to customer needs at the right cost and also within stipulated time.

Some popular casting simulation softwares which are available to foundry engineers are AutoCAST, MAGMASoft, ProCAST, SOLIDCast, CAP/WRAFTS, CastCAE, Castflow, Castherm, JSCast, MAVIS, Nova-Solid/Flow, PAM-CAST, RAPID/CAST, and SIMTEC etc. These simulation software generally follow any one the following numerical technique to solve the related differential equations, Finite Differences Method (FDM), Finite Volumes Method (FVM), Finite Element Method (FEM) and Vector Element Method (VEM) etc. ProCAST, SOLID Cast, OPTI Cast and FLOW Cast are based on the FEM technique while QuikCAST is based on FDM technique and AutoCAST is based on VEM technique.

It is easily understandable that the simulation software will perform well if and only if the input parametrs are close to the real world values. Most critical input values for such simulation software are the thermo-physical properties of cast metal and mold, as well as interface boundary conditions. But these values are temperature dependent. Therefore, the values are difficult to acquire for different metal-mold-process combinations. As a result, the outcome of simulation software may deviate from reality.

In this chapter, advantages of casting simulation software, limitations, and some best practices are illustrated. In addition, some live examples have been cited to understand the process logically and scientifically.

2. Application of casting simulation software

Casting troubleshooting, method optimization and part design improvement are the main three applications of casting simulation software. Therefore, these are described here in brief.

2.1 Casting Troubleshooting

As the title suggests, trouble in existing casting like high or varying level of internal defects (shrinkage porosity, sand inclusions, cold shuts, etc.), or poor yield of castings can be eliminated by using the simulation software. To address such problems, casting engineers' first use to calibrate the software by making the exact platform in the virtual world. Now the simulation defect and real world defects are compared for calibration. Once achieved, the engineer can change the inadequate size or location of feeder or gate. It is also a fact that the simulation software reveals the defect positions where otherwise in general engineer don't look for. Regarding the improvement of yield, the foundry engineer can play with the simulation software in a hope to reduce the oversized feeders or risers (if any). If no defect is observed in simulation software by reducing the feeder dimensions then the engineer can go for shop floor trial. If the result agrees then definitely, the yield will improve.

2.2 Method optimization

In methoding of castings, solidification simulation executes an essential role in acquiring the best possible quality of castings. Simulation software is beneficial for both under development as well as existing castings, and reduces number of shop-floor trials. The methodings for example, casting orientation, mould layout, feeders, feed-aids, and gating etc. are altered on a computer, and pretended to review for defects, if any. Numerous iterations are performed until the coveted quality and yield are obtained. It is noteworthy to mention that even trivial, insignificant advancements in existing castings that are being manufactured in huge numbers, can direct to notable enhancements in the utilization of matter, energy, machinery and labour resources. Likewise, simulation is decisive for large, heavy castings under development since the cost of trials or repair is limited. Several programs have been included in the algorithms for automated (user-guided) optimization of feeders and gating channels [3]. Many researchers have proclaimed that approximately 90% of the defects in parts are due to errors in design and hardly 10% are due to production difficulties. The casting simulation software can also be used for calculating the cost of the job in an indirect manner. In the very initial step of design of a to be cast part should be simulated, otherwise it may so happen that the proposed part is not at all castable. A thumb rule says that the cost to change in design increases ten times in every step of the design and manufacturing process. Therefore, method optimization should be done using the casting simulation software as early as possible in a designing process and this practice can save a lot of money both in foundry and machine shop.

2.3 Part design improvement

Thick junction and long thin section in a casting may result in shrinkage porosity and cold shut. Therefore, if a part designer encounters such situations then he should immediately consult the foundry engineer to check the castability of the part using casting simulation software. Early detection of castability may also insist the part designer to do minor change without affecting the functionality of the part.

3. Operational Methodology of simulation software

This section has been discussed in following three subsections.

3.1 Inputs

Maximum commercially available casting simulation software are Finite Element Method based [4]. So this discussion is applicable for FEM based simulation software.

- 3D CAD model is the main input for any casting simulation software. The CAD model can be created using a solid modeling program. Model of the part can be obtained from OEM customer.

- Various allowances like draft, machining, shrinkage, distortion etc. are to be provided on the CAD model if not given earlier.

- Now, the model has to mesh. Meshing means that the model is to be split into several simple elements. A tinier mesh size returns more delayed but more stable outcomes. Adaptive meshing (finer in decisive domains and coarse elsewhere) provides quicker outcomes without compromising on the precision.

- Following meshing, material characteristics like density, thermal conductivity, specific heat, latent heat etc. are to be provided as input to the software.

- Next, the boundary conditions have to be defined.

- Simulation software furnishes reliable and precise outcomes if the CAD model, FEM mesh, material characteristics and boundary conditions are exact.

3.2 Outputs

- The principal outputs of simulation applications incorporate animated visualization of mould filling, casting solidification, and further cooling to room temperature.

- Mould filling simulation assists in forecasting the total filling time as well as help in predicting following casting defects like mould erosion (heading to sand inclusions), incomplete filling (cold shuts and misruns), and air entrapment. Blowholes, produced by entrapment of gases, are yet difficult to predict.

- The outcomes of solidification simulation incorporate colour-coded freezing profiles with respect to time. These temperatures profiles help in predicting the position of shrinkage porosity based on Niyama criteria.

- The casting simulation software can also predict the microstructure, mechanical properties, residual stresses etc.

3.3 Analysis

Casting simulation software can only help foundry engineers to analyze the effect of a particular method design on the yield and quality. The software enables us to 'look through' the virtual mould. However, it can not perform better by itself. The outcomes should be analyzed by a knowledgeable foundry engineer. Application of casting simulation software can improve the productivity of a company and also it is a tool for the foundrymen to succeed but it cannot substitute him.

4. Intelligent design assistant

As stated earlier, Simulation software furnishes reliable and precise outcomes if the CAD model, FEM mesh, material characteristics and boundary conditions are exact (otherwise: garbage in, garbage out). Material characteristic and boundary conditions data usually have to be ascertained and fine-tuned by experimentation. This exercise may take numerous weeks, which is beyond the scope of average organizations. Simulation programs demand engineers with higher educational qualifications, CAD/CAM experiences and casting design knowledge to conduct the simulation and interpret the outcomes correctly. The programs are computation-intensive and need robust engineering workstations. Even then, any particular iteration of CAD model making, mesh creation, boundary condition stipulation, simulation and visualization can consume 2–5 days for an intricate component. Thus it may demand many days to reach an optimal casting scheme.

To address these issues now a days, some simulation software provides a single integrated environment for casting design, modeling, simulation, analysis and project data management. Advanced geometric reasoning and knowledge-based functions have been incorporated in the software, making it work like an intelligent assistant to casting engineers [5]. The methodology for casting design (mainly feeding and gating systems), process simulation, castability analysis and optimization is performed by intelligent simulation software is explained below.

- The CAD model of the part is exported in standard. STL format and imported into the simulation software.
- The program automatically recommends the mould size and subsequently upon the approval from the user, generates the mould model encompassing the part model.
- Now, an initial casting solidification simulation was performed.
- The program automatically produces the mesh, fixes the boundary conditions, measures the advancement of solidification, post-processes the results and demonstrates the position and degree of shrinkage porosity. All the events, as mentioned above, take less than 15 minutes on a Pentium computer.
- Now if any porosity is observed in the simulation results then a chill is modeled automatically in that zone to increase the heat transfer rate. Modified simulation reveals that there is reduction in porosity.
- Likewise, the simulation software automatically calculates modulus (ratio of volume of the casting to the surface area of the casting) in the hot spot zone of the casting and properly designs the feeder dimensions in such a way that its solidification time is more than the hot spot. After user approval, the feeder model is built automatically.
- Additionally, the gating plan is designed by this program semi-automatically. The user exclusively defines the ingate joining points on the part surface. The program automatically proposes the sprue location and the runner path, which can be altered by the user if needed. Then the dimensions of all gating parts are automatically calculated, and a solid model of the gating system is built.
- The solid models of the feeder, chill and gating system, are represented for visual feedback. The size of feeder and chill are optimized through several iterations of design-model-simulate-analyze until simulation prognosticates zero porosity defects even for the highest quality requirements.
- Finally, casting is designed on a scale of 0-100. A value of zero implies impossible to cast and 100 indicates ideal castability. The actual values usually lie in-between.

5. Example: workflow of a casting simulation software (Z-CAST)

The Z-CAST is Finite Difference Method (FDM) based simulation software. To create a new project following steps are to be followed.

- The 3D model in. STL format is required to be imported
- Virtual mould is to be created
- Meshing is done. It is dependent on three parameters, Geometry complexity, Minimum wall thickness and Weight of the component.
- Material and initial temperature are to be provided
- Ingate of the casting is to be designed
- To obtain the graphcal representation of simulation behavior, some virtual thermo-couples are to be set on some strategic locations
- Next, the solver environment is to be declared. Here some typical inputs are required from users. For example, the analysis terminate conditions are flow rate and time are to be declared. Input wall condition is to be provided. For this, there may be two situations like slip condition and no-slip condition and hence the input value will vary between 0-1 respectively. For sand casting input wall value is between 0.4 to 0.6 whereas for die casting the value is 1. Next, the riser material type, whether it is exothermic or not is to be given as input. Heat transfer coefficient is to be provided thereafter.
- Once done, the software will provide the results with flow temperature, solid time, solid temperature, details on shrinkage condition etc.

6. Case study 1

For preparing some wedge blocks, a scheme of preparation mechanism used by a manufacturing company is shown in **Figure 1**. The product being manufactured with the existing method design contains shrinkage defect. The simulation result also confirms the same. The shrinkage defect and its associated simulation result are shown in **Figure 2**. Therefore to eliminate the defect and also improve overall process associated with the production stack moulding is adopted as shown in **Figure 3**. Due to this modification various related parameters also improve as shown in **Table 1**. The techno-economy analysis of wedge casting per ton shows that cost per piece reduces and productivity increases with stack moulding approach.

7. Case study 2

The interconnection of coaches by a coupling makes a train. In earlier days screw coupling was used. Screw coupling has certain inherent limitations like haulage of longer train is not possible in freight, climbing of coaches in collisions and derailment, life of shunting staff at risk and higher maintenance staff requirement. On the other hand, central buffer coupling (CBC) has advantages like the coupling is safe for shunting staff, less time is required since quick detachment is possible, less staff for uncoupling, the coaches do not climb on each other during accident and hence prevent damage to life and property during accident. The CBC has three major components, knuckle, coupler body and yoke. The CBC is a cast product. In this present case study the use of casting simulation software on a CBC product is demonstrated. The simulation software is capable of indicating the loop-holes of

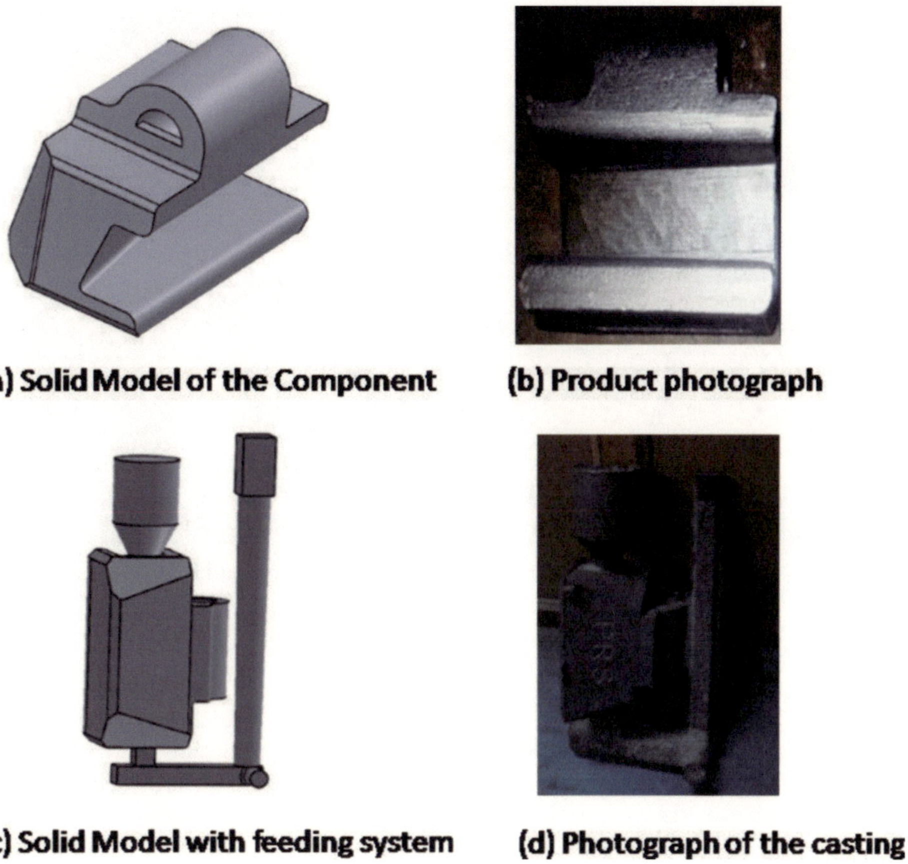

Figure 1.
Scheme of wedge block preparation

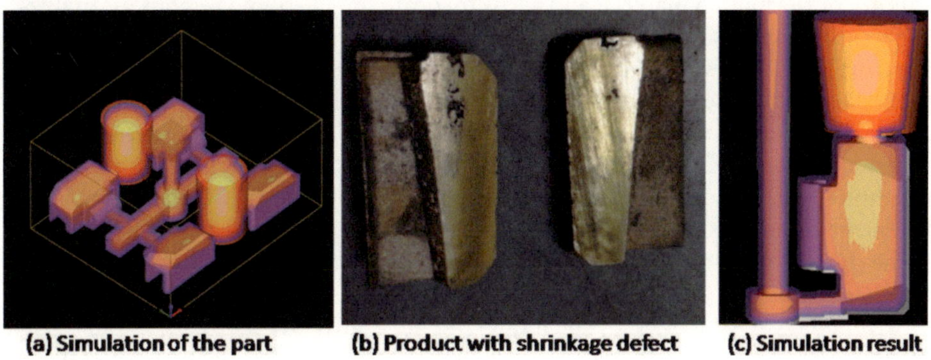

Figure 2.
Product with shrinkage defect and linked simulation

the existing design and the caveats can be ironed out by proper methoding. Final simulation results show that there is hardly any detrimental defect in the cast part and thus a huge cost for trial and error is saved. To demonstrate the role of methoding and solidification criterion on casting defects, one crucial industrial case study is discussed here.

The foundry is producing three major components of CBC i.e. coupler body, knuckle and yoke in a single mould. Cast Steel (ASTM M-211 GRADE-E) was used as the casting material and green sand under high pressure moulding system was

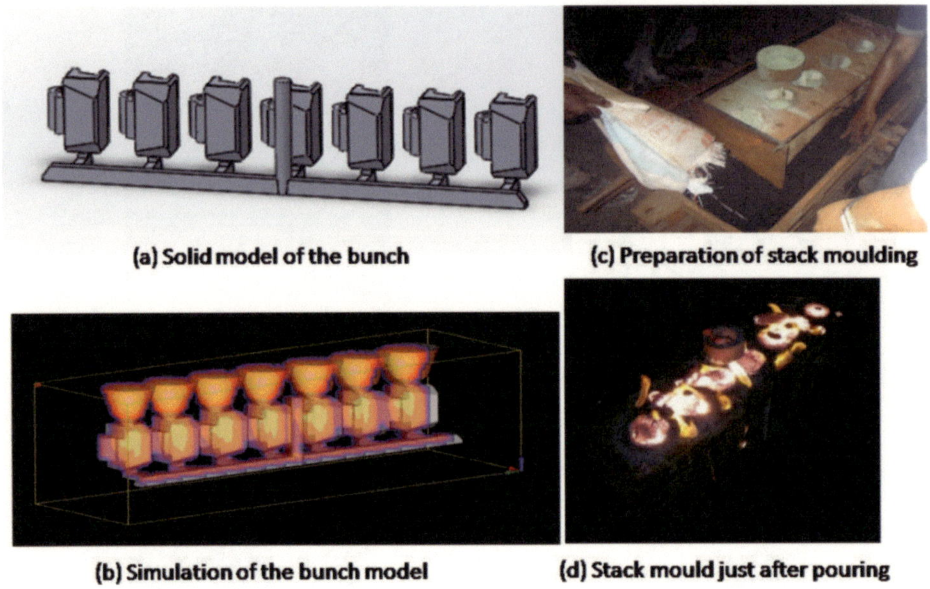

Figure 3.
Scheme to indicate the modification done

used. The 3D CAD model is created in solid modeling software and converted to .STL format. The .STL file imported into the Z-CAST simulation software. Here, pouring temperature is taken as 1610 °C, and pouring time is 30 seconds. Exothermic sleeves are used in this case and the mould temperature is considered as 30°C before the pouring of molten metal. The shrinkage allowance of the material is also considered as 3.5%.

The flow simulation of the product shows filling percentage with respective temperature scales as shown in **Figure 4**. The flow temperature analysis can test fluidity of the molten material and assist to locate the un-filling regions by analyzing temperature plots in thin sections of the cavity. The white colour in the temperature scale corresponds to the highest temperature of the melt i.e. the pouring temperature. Yellow, red, dark blue and finally light blue are indicating temperatures of gradual decreasing order. Filling condition of this product at different filled position (2, 10, 40, 60, 80 and 100%) is shown in the figure. **Figure 4a** shows that the molten metal is entering in the mould cavity through sprue. Mould filling is an essential matter since the melt temperature will commence decreasing when it comes in contact with the cooler mould surface. **Figure 4a** shows that molten metal will flow according the slope of mould cavity. **Figure 4b** shows that 10% volume of the mould cavity is filled by the molten metal. **Figure 4(c-e)** indicate that 40, 60 and 80% space of the mould cavity is filled by the melt and the liquid metal has started solidification process. **Figure 4f** shows that at 100% filled condition the components at the farthest point from the sprue have blue colour which indicate that the temperature at those points are sufficient lower than the pouring temperature. At the end of 30 seconds i.e. the complete pouring of metal, temperature drops from 1610°C to 1490°C

The air entrapment simulation is shown in **Figure 5**. This flow un-filling simulation shows when molten material enters into the mould cavity, how the entrapped air escapes and thus this simulation can predict the unfilled locations of the cavity. Here, the red colour indicates cavity yet to be filled and transparent area shows metal is filled. Small dots indicate the amount of entrapped air in the cavity.

Factors Considered	Conventional	Stack Moulding
Bunch Weight	42.96 kg	75.19 kg
Weight/piece	5.80 kg	5.80 kg
No. of pieces/bunch	4 nos	7 nos
Net Casting weight/bunch	23.20 kg	40.60 kg
Bunch Yield %	54%	54%
Metal Cost	22644.93.(INR)	22644.93 (INR)
Ferro Alloy Cost	2485.40 (INR)	2336.27 (INR)
Electricity Cost	15395.74 (INR)	14472.00 (INR)
Core cost/Molding cost	1640.24 + 6481.00 (INR)	16223.82 (INR)
Store Cost	7408.04 (INR)	6768.32 (INR)
Heat Treatment Cost	3500.00 (INR)	3500.00 (INR)
Shot Blasting Cost	227.00 (INR)	227.00 (INR)
Labor Cost (Production)	4887.45 (INR)	2625.26 (INR)
Labor Cost (Fettling)	2930.00 (INR)	2758.00(INR)
Maintenance Cost	1300.00 (INR)	1300.00 (INR)
Total Variable Cost	68899.80 (INR)	72855.60 (INR)
Fixed Cost	4200.00 (INR)	4200.00 (INR)
Rework	7000.00 (INR)	500.00 (INR)
Total Cost	80,099.80 (INR)	77,555.60 (INR)
Cost per piece	570.57 (INR)	561.55 (INR)
Quality of the casting Rejection	Good 30%	Very Good (ASTM 2) 5%
Aesthetic Look	Standard	World class
Processing time	1 Unit	1/3 Unit
Consistency	Not up to the mark	Very Consistent

Table 1.
Techno-Economy Analysis of Wedge Casting/Ton

Figure 5(a) shows the level of entrapped air at a level of 20% filled mould cavity while **Figure 5(b)** shows when the metal pouring has been completed. The final picture depicts that no air particles have been entrapped observed in cavity during filling.

Solidification temperature pattern shows rate of solidification and time taken to solidify. It is evident from left top corner of **Figure 6(a)** shows after elapse of 123 seconds the solidification is 1%. Similarly **Figure 6(b)** indicates that it takes 2910.558 seconds to solidify 100%. The temperature scale shows that some portion of the product has been delinked with the riser and contains higher heat. This will cause shrinkage defect in those locations. The yellow arrows of **Figure 7** indicates the major shrinkage cavities produced after solid simulation and these hot spots are needed to be taken care of.

Now, the detail component wise defect analyses of the knuckle, coupler body and yoke have been shown in **Figures 8**, **9** and **10** respectively. **Figure 8** shows that the knuckle is free from shrinkage, blow holes and air entrapment issues. As per Research Designs and Standards Organization (RDSO) standard, there is no porosity

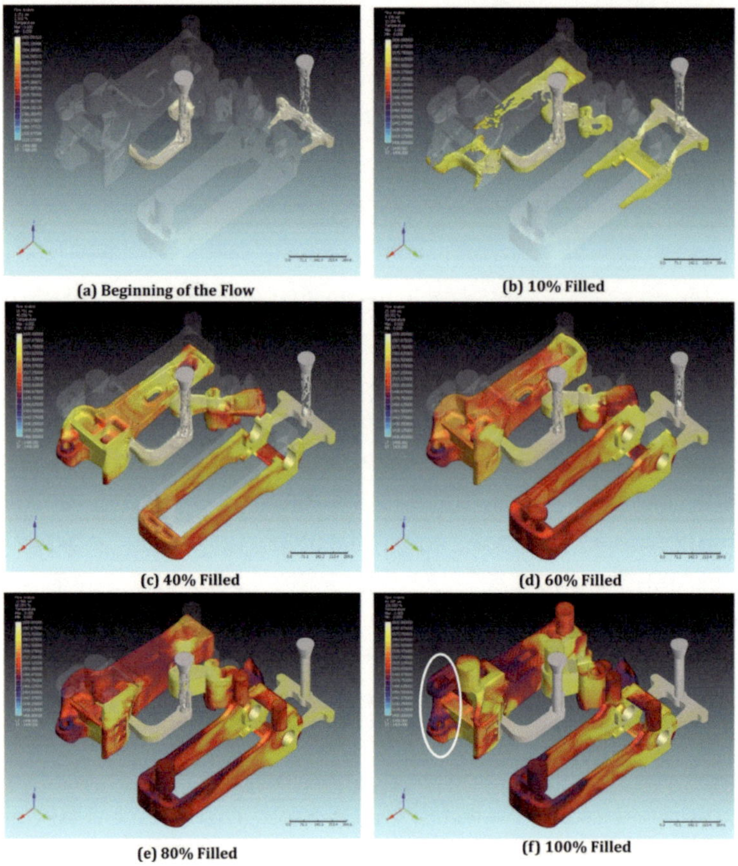

Figure 4.
Simulated filling conditions at different filled positions

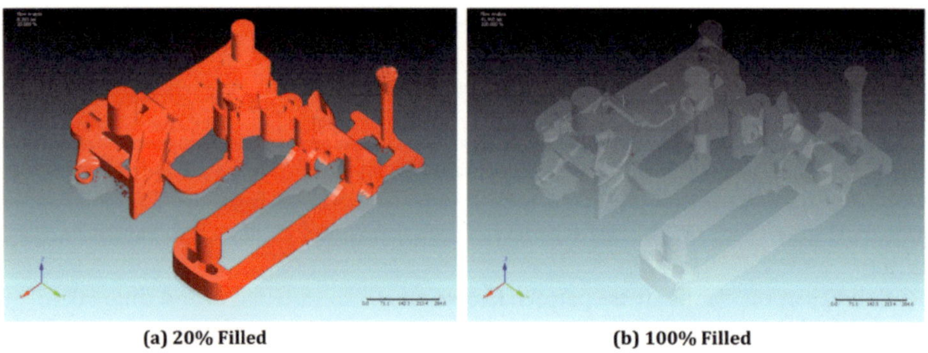

Figure 5.
Air entrapment simulation

in critical location of the knuckle component. But in other non-critical locations some porosity can be observed but those would be passed through the radiography NDT testing of level 2.

Figure 9 indicates that major shrinkage defects have formed in the coupler body and these are to be removed by adopting proper casting methodology. The figure also indicates some major porosity formation which are crossing the level 2 of radiography testing, therefore, these porosities are also needed to be eliminated

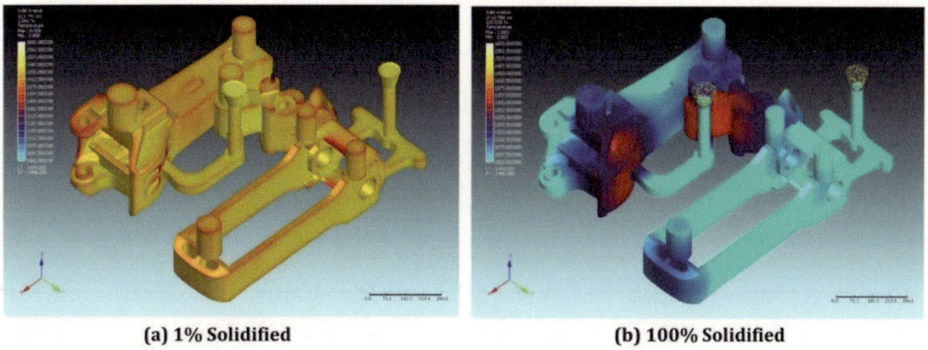

Figure 6.
Solid simulation of product

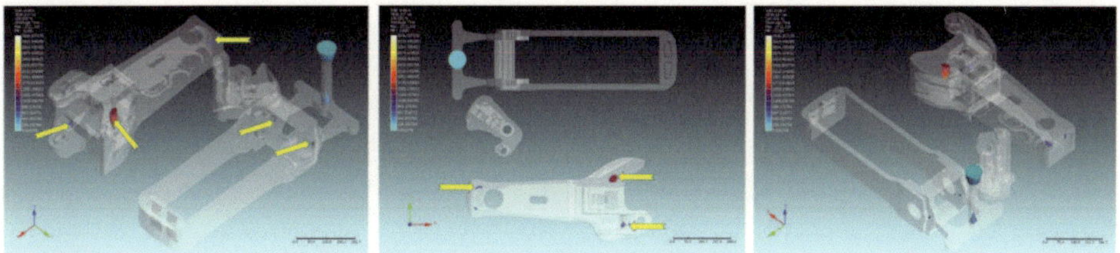

Figure 7.
Indicating shrinkage cavities after solid simulation

Figure 8.
Defect analysis of the knuckle and real cross section of critical location indicating zero defect

from the cavity. In some non-critical locations, some minor porosity has formed but these are under level 2.

While analyzing **Figure 10**, it is found that the yoke is free from major shrinkage defect but some minor shrinkage defects have been observed (indicated by yellow

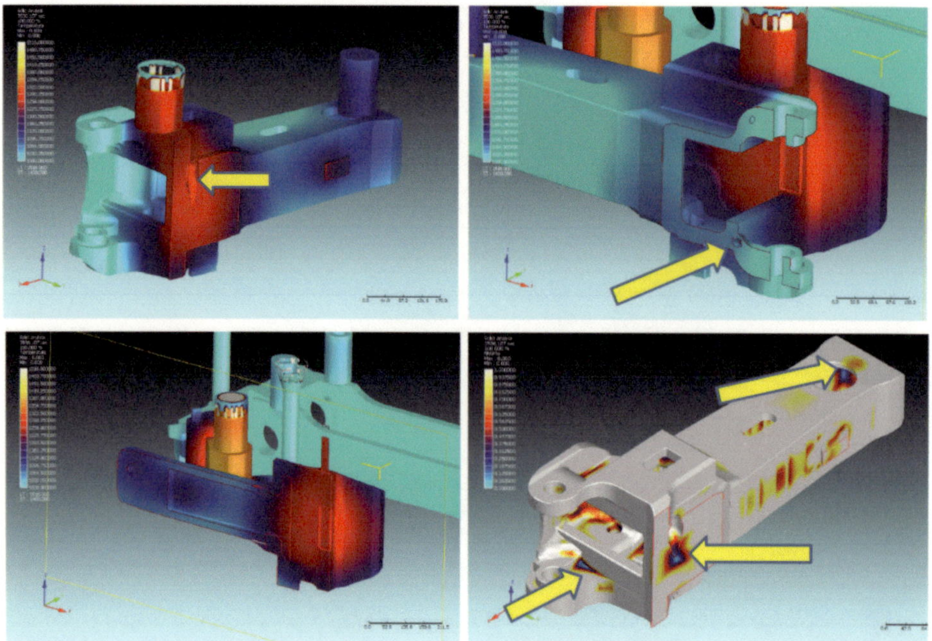

Figure 9.
Defect analysis of the coupler body

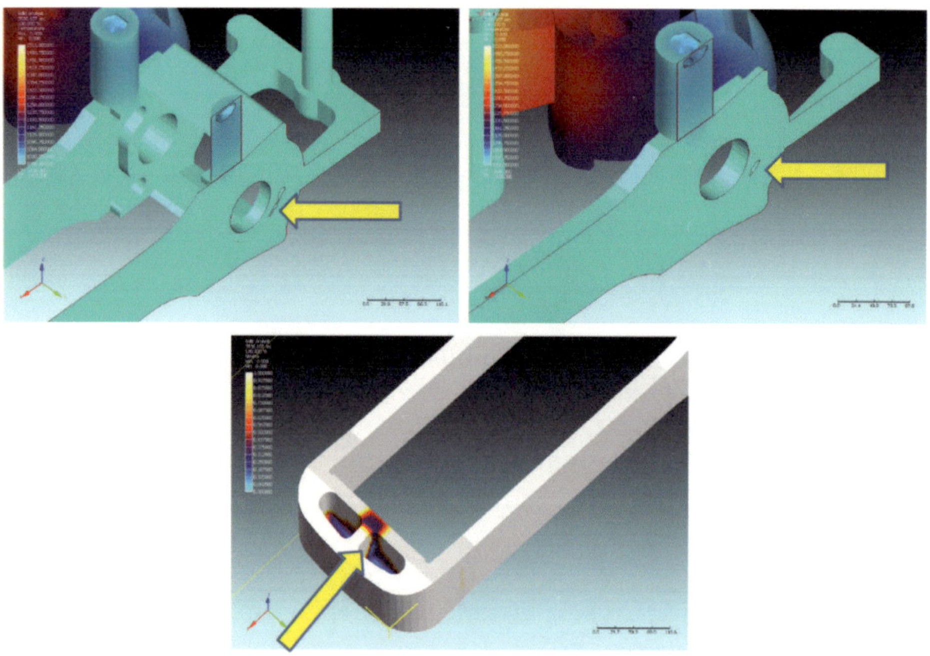

Figure 10.
Defect analysis of the yoke

arrows) along with some minor porosity below the riser, which needs to be eliminated from casting. The porosities crosses the radiography level 2 and therefore, needs to be eliminated from the cavity.

7.1 Iteration

By analyzing **Figures 4–10**, it can be concluded that knuckle has no major issues to be solved but the coupler body mouth and one end of yoke are prone to shrinkage defect and porosity respectively. Cross-section of critical section before modified methoding is shown in **Figure 11**. To solve the issues, two sprue with existing gating system is used with following minor modifications.

a. Minor modification in riser sizes: If a close comparison is done between **Figure 6(a)** and **Figure 12(a)**, it can be found that the dimensions of the risers are different. It is needless to mention that the authors are unable to share such data in detail.

b. In mouth area of coupler body, chromite sand core (shown as the red arrow in **Figure 13(a)** and the schematic of the core shape is shown in detail in **Figure 13(b)**) is to be used to have effect of chill in order to minimize the major hot spot in the critical location. **Figure 9** indicates that the mouth of the coupler body is most prone to shrinkage defect and to eliminate that use of chromite sand core is necessary since it has high thermal conductivity and good chilling effect.

c. The yoke is also prone to shrinkage and porosity defect as discussed in **Figure 10**. To eliminate this defect, a riser should be used as shown by a brown arrow in **Figure 13(a)**.

d. Additional three more chills are required and suggested to be placed in the mouth core of the coupler body else a small riser should be incorporated. The use of chills or a riser is technically equivalent but both chill and riser have certain pros and cons. The use of extra riser is associated with some extra cost related to material, energy to melt, cutting, re-melting etc. Whereas use of chills does not have such extra cost, they are to be purchased or fabricated once. But in practical shop floor, it is seen that after several production runs, the labours forgot to collect all the chills while breaking the sand mould and eventually the chills are misplaced. Therefore, it becomes a decision of the proprietor of the company, whether he will go for chill or a riser. In this case a riser is used which is marked as a deep blue arrow in **Figure 13(a)**.

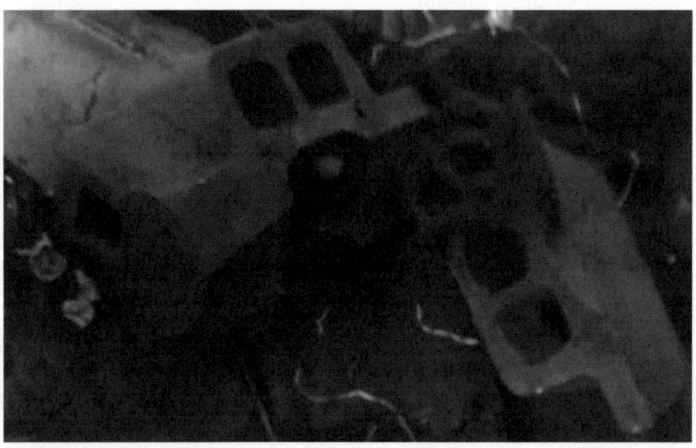

Figure 11.
Cross-section of Critical Section before modified methoding

e. Some ventings (shown by black arrow in **Figure 13(a)**) are suggested to increase permeability. In this case the mould is made using high pressure moulding so the permeability of the mould is very low, and making of manual conventional vent just before metal pouring is very tough in such cases. So while, making the mould cavity, some rods f small diameter are attached with the pattern strategically.

f. Exothermic powder is also advised to be used immediately after pouring the metal to minimize heat loss.

Figure 12(a) clearly indicates that the total solidification time is 3869 seconds i.e. 958 seconds (Approximate 16 minutes) more time than the solidification time

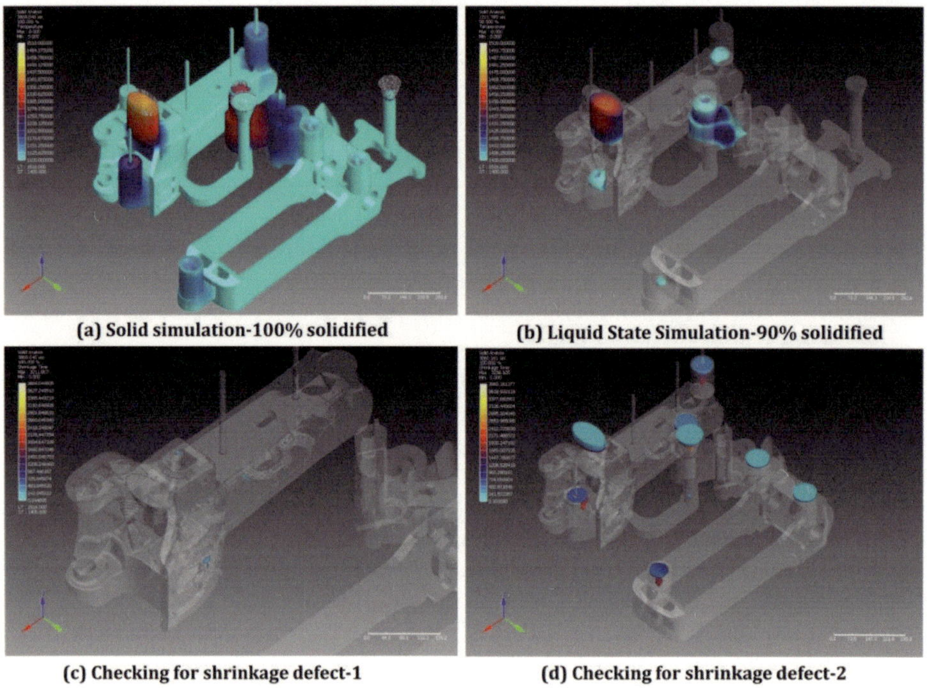

Figure 12.
Defect analysis of the product after modifications

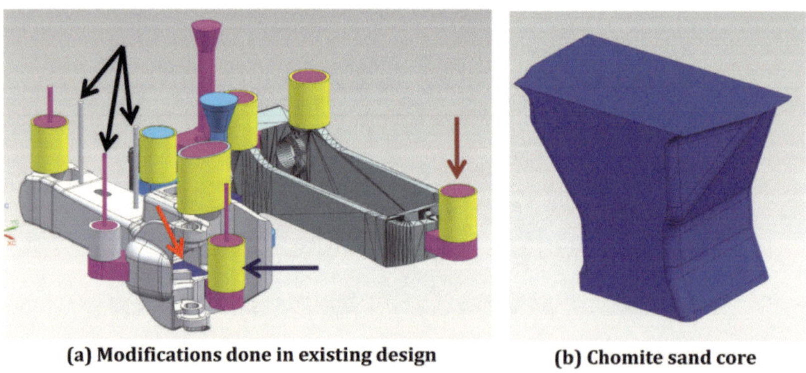

Figure 13.
Modifications done in the existing design to obtain sound casting

consumed before modification. Liquid state simulation of **Figure 12(b)** shows that a smooth directional solidification has been achieved and all major hot spots have been eliminated from the cast components. **Figure 12(c-d)** indicates occurrence of minor shrinkage defect in the casting but these are not detrimental to the component.

8. Conclusions

In this study, the advantages of casting simulation software, limitations, and some best practices have been observed along with some case studies to understand the process logically and scientifically. Simulation technology has appeared as a blessing to foundry engineers to implement virtual experimentations, forecast casting defects and enhance the existing casting design without melting metal in furnace. This not only reduces the expense of die modification and material/energy costs but also contributes a more beneficial insight into the process. On the other hand, with the advancement of technology, to produce zero defect castings with improved yield, it is very much required to optimize the casting design (feeding and gating). This goal can be accomplished by coupling intelligent design assistant module of simulation software. This will enable us to complete the design-simulate-analyze cycle within a time less than one hour. This tool enables us to optimize complex castings in a single day. This also assures more acceptable casting design, quicker than ever before.

Acknowledgements

Authors wish to convey their heartfelt thanks to Mr. Prashant K, Senior Ex-Application Manager, DHIO research and engineering private limited for his support in while making the Z-Cast simulations for this case study. The authors are also indebted to D N Ganesh Babu, General Manager, Product Sales & Services, DHIO Research & Engineering Pvt. Ltd. for his continuous support for giving the Z-Cast software. Author also thankfully acknowledge the contribution of Mr. Abresh Ranjan Das for his continuous support & help for sharing his profound experiences to complete this work unless otherwise this work could not get the present shape.

Conflict of interest

The authors declare no conflict of interest.

Author details

Partha Haldar[1] and Goutam Sutradhar[2]*

1 Department of Mechanical Engineering, Government College of Engineering and Ceramic Technology, Kolkata-700 010, West Bengal, India

2 National Institute of Technology, Langol, Imphal-795004, Manipur, India

*Address all correspondence to: cast_1963@rediffmail.com

References

[1] Ravi B. Computer-Aided Casting-Past. Present and Future. Indian Foundry Journal. 1999;**45**(1):65-74

[2] Ravi B: Casting Simulations - Best Practices. Transactions of 58th Indian Foundry Congress. 2010.

[3] Ravi B, Joshi D: Feedability Analysis and Optimisation Driven by Casting Simulation, Indian Foundry Journal. 2007: 53(6): 71-78.

[4] Lewis RW, Huang HC, Usmani AS, Tadayon MR. Solidification in castings by finite element method. Materials science and technology. 1990;*6*(5):482-490

[5] Ravi B. Computer Aided Design And Analysis For Zero Defects. In: International Conference on Aluminium; December 2000; p. 1-6.

Chapter 5

CFD Optimization Method to Design Foam Residue Traps for Full Mold Casting

Yuto Takagi, Masahiro Inagaki and Ken'ichi Yano

Abstract

Full mold casting is a casting process in which a mold made of wood or metal is substituted for a styrofoam model. This metal casting process is advantageous for the production of large-sized castings because it uses a foamed model. However, this unique process of melting a foamed model causes a problem which is the foamed model remains dissolved in the casting. This is called foam residue defect and is specific to full mold casting. In this study, we propose a new casting design called a residue trap to reduce these residue defects. This residue trap collects the residue of foam models included in the molten metal, which tends to be generated when the temperature of the molten metal becomes low by being attached to the product part in the same way to overflows. We also optimized the shape of the residue trap in terms of easing of post-treatment and increasing efficiency of collecting foam residue. Eventually, the effectiveness of the residue trap was verified by actual full mold casting experiments.

Keywords: full mold casting, foam residue, CFD simulation, shape optimization, genetic algorithm

1. Introduction

Unlike conventional sand-casting using wooden molds, full mold casting offers the advantage of near-net shaping without the need for a draft angle in the product shape. As a result, the number of person-hours in post-casting processing can be decreased, reducing cost and shortening delivery times. However, the occurrence of residue defects, which is a phenomenon unique to the full mold casting method, is a major problem in the manufacture of large castings by this method. This is a serious issue that significantly impairs the quality of the product because the unresolved part of the foam model is mixed in with the product.

As a measure against defects in residue at the casting site, the pouring temperature can be increased to facilitate the thermal decomposition of a foam model. However, due to the high temperature inside the product, the volume shrinkage due to solidification increases and shrinkage nests tend to occur. On the other hand, if the pouring temperature is lowered to prevent shrinkage cavity, the foam model cannot be thermally decomposed efficiently and residue defects are generated. Although there is also a method of installing multiple gates, which are the inlets to the product, to improve the molten metal turnover, a flow of metal that rolls in the foam model is generated, and residue defects are more likely to occur.

Traditionally, these casting conditions and schemes have been determined based on the experience and intuition of skilled workers. In recent years, however, the use of computational fluid dynamics (CFD)-based simulations has made it possible to study appropriate conditions and solutions in advance based on the analytical results of the melt flow and solidification processes, thereby reducing the cost and time required for trial manufacture and experimentation. Particularly, solidification simulations are used to study casting conditions and solutions for shrinkage cavity defects, which can be estimated with high accuracy by comparison and verification with experiments [1]. The authors have also been optimizing the shapes of the runners in die castings by using CFD simulation and the shape optimization method for molten metal flow [2].

In recent years, the full mold casting method and the vanishing model casting method have been actively researched. Maruyama et al. have collected, analyzed, and investigated the pyrolysis products generated during the filling of molten metal in the vanishing model casting process [3]. Koroyasu et al. reported on the adiabatic properties of a coating in the vanishing model casting method and conducted a simple simulation of molten aluminum alloy flow in that method to investigate the influence of the air permeability of the coating, the presence or absence of depressurization, and casting methods on molten metal flows [4]. Karimian et al. investigated the influences of the coating thickness on product quality. When the coating film is thin, the amount of pyrolysis gas emitted from the foamed model is higher, and the amount of residue due to pyrolysis is reduced [5]. In addition, studies on the analysis of molten metal flow behavior in vanishing model casting have been actively conducted using CFD simulation [6, 7].

This study proposes a new casting design for the production of large castings by full mold casting. In this method, a residue trap is installed at the product part to prevent the occurrence of residue defects, and the optimal design of the residue trap is realized by using CFD simulation and the shape optimization method. The effectiveness of this method is demonstrated through casting experiments using an actual full mold casting.

2. Parameter identification for CFD simulation

In general castings such as sand casting, molten metal is poured into a hollow mold. In contrast, in full mold casting, molten metal replaces a foam model. Therefore, because the flow of metal differs greatly between full mold casting and normal casting, it is necessary to understand the phenomena that actually occur in the mold.

In full mold casting, the heat of the molten metal decomposes the foam model and generates gaseous pyrolysis products. The thermal decomposition products are successively discharged through the coating mold into the sand mold. If the gas layer of the pyrolysis products is thin, the rate of heat transfer from the molten metal to the foam model will increase, as will the rate of molten metal flow. Conversely, the thicker the gas layer of the pyrolysis products, the lower the heat transfer rate and the lower the flow rate of molten metal. Therefore, the flow metal is determined by the complex interaction of multiple factors such as the metal's temperature, its position in the mold, the material and morphology of the foam model, and the thickness and composition of the coating mold.

In the CFD software used in this study, a foam model is represented as an obstacle that disappears when heated. Therefore, the flow of molten metal is hardly affected by pressure and inertia, but depends on the heat transfer phenomenon. Specifically, the heat transferred from the metal to the foam model at their interface is calculated. Subsequently, depending on the amount of heat, the foaming model disappears and the metal migrates. Therefore, the average inflow velocity U_{front} of the interface between the molten metal and the foam model can be approximated by the following Equation [8].

$$U_{\text{front}} \approx \frac{H_0}{\rho C_{\text{p}}} \tag{1}$$

where H_0 is the apparent heat transfer coefficient between the metal and the foam model and represents the ease of heat transfer to the foam model. The products of ρ and C_{p} are the density and specific heat of the foam model.

In the full mold casting, the heat of the metal decomposes the foam model and generates gaseous pyrolysis products. The pyrolysis products are discharged into the sand mold through the coating process. These products are smaller than the metal when the metal is above the foam model. Therefore, the thickness of the gas layer between the metal and the foam model due to the pyrolysis products is reduced. Therefore, the heat transfer rate is expected to be high. On the other hand, when the metal is located below the foaming model, the thickness of the gas layer between the metal and the foaming model is increased and the heat transfer rate is considered to be low. In the CFD simulator used in this study, considering this effect in the gravity direction, the flow rate of metal in the gravity direction is defined by Eqs. (2) and (3) using the parameter E_{g} [8].

$$U_{\text{flont}} \approx \frac{H}{\rho C_p} \tag{2}$$

$$H = H_0 \cdot \left(1 + E_g \cdot sign(g_n) \cdot \frac{\sqrt{|g_n| \cdot l}}{\sqrt{|g_n| \cdot l} + \frac{H_0}{\rho C_p}} \right) \tag{3}$$

where g_{n} is the perpendicular component of the gravitational acceleration to the metal interface and l is the surface roughness of the metal. The product of these

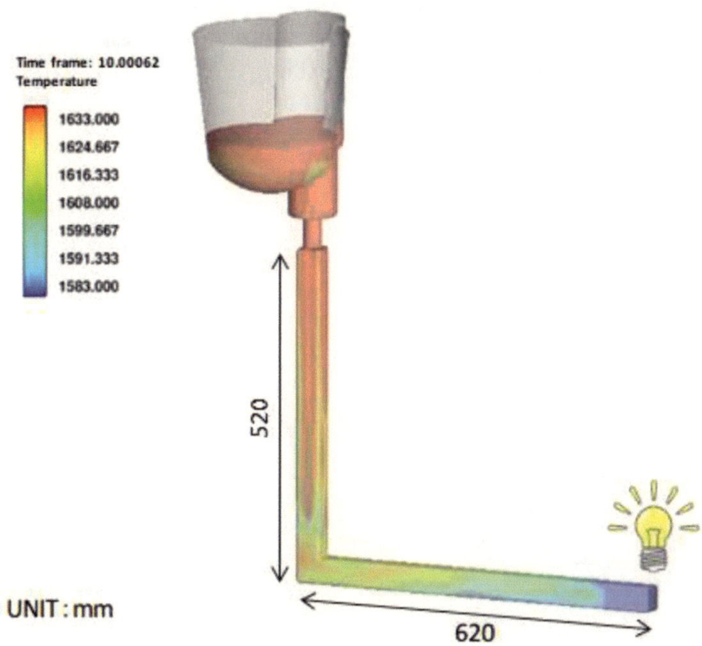

Figure 1.
CFD simulation model for identification experiment.

values expresses the difficulty of heat dissipation of the foam model. Although the model expressed by this equation does not exactly represent the generation of pyrolysis products and the thickness of the gas layer, it is thought that the flow of metal, which is close to the real phenomenon can be expressed by setting appropriate values for each parameter. In this study, casting experiments and molten metal flow analysis were conducted for the model shown in **Figure 1** to obtain the appropriate values of the unknown parameters H_0 and E_g in Eq. (3).

The metal was FC300 gray cast iron, and the pouring temperature was set at 1683 K. Foamed polystyrene with a foaming factor of 60 times was used for the model. A water-based mold coating agent for full-molded cast iron was used for the coating, and the coating film was about 2 mm thick. Silica sand with an AFS grain size index of 36.7 was used as the casting sand. A stopper system was used for pouring the metal. The arrival time of the metal was measured by inserting a touch sensor into the end of the model; this sensor emitted. The experimental results show that the start to the finish of pouring was 7.2 s. The casting weight was 35 kg.

Next, CFD simulations were performed. The parameters in Eq. (3) were set to ρ = 16.7 kg/m^3 and Cp = 2100 J/(kg-K). In order to make the arrival time of the metal consistent with that of the experiment and the simulation, the parameter H0 was adjusted based on the definition of Eq. (3), and good results were obtained at H0 = 5500 W/(m2-K) and Eg = 0.5. We therefore use these values in the simulations in this study.

3. CFD analysis and new gating system

3.1 Design of gating system based on CFD analysis

In this study, a capstan drum (**Figure 2**) is used as the product shape. Also, a casting runner with a diameter of 30 mm is used for the solution. The capstan drum was analyzed for metal flow using the two gating systems shown in **Figure 2(a)** and **(b)**.

Gating system A has a gate directly on the arm, which is considered to be prone to many residue defects. Since an enormous amount of time would be required to analyze everything from the pouring to the filling of the product part, a simple

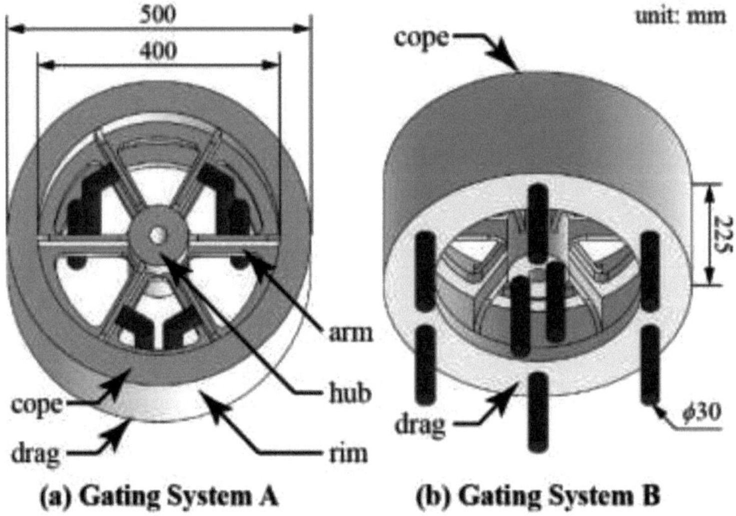

Figure 2.
Shape of capstan drum and gating systems.

model in which metal flows into the product part at a constant pressure from the gate was used. The inflow pressure was set at 1.425×10^5 Pa in absolute pressure, and the temperature of the pouring molten metal was set at 1623 K. The pouring temperature of the metal was set at 1.425×10^5 Pa in absolute pressure. In addition, since the shape was symmetrical, we were able to cut the analysis area in half to reduce the analysis time.

The results of the CFD analysis are shown in **Figure 3(a)**. As shown by the dashed ellipses, the metal flows by gravity from top to bottom. Due to the difference between the density of the metal and the density of the gas and residue generated by the pyrolysis, the metal would entrain the gas and residue, which would affect the occurrence of residue defects.

Then, gating system B is modified to install gates at the rim and hub to improve the flow of gating system A. To prevent the formation of defects in the residue, the pouring temperature was increased by 20 K to 1643 K.

A molten metal flow analysis was performed on the drum using this gating system B. The results of the analysis are shown in **Figure 3(b)**. The flow in the rim proceeds gradually from the bottom to the top, which makes it difficult for the metal to entrain gas and residue. In addition to the increase in temperature, these flows may actually reduce the number of residue defects.

In other words, if a gate is installed on the arm as in gating system A, a drop-in flow occurs, leading to the generation of residue defects, so the gate must be installed on the bottom of the hub and rim as in gating system B. If the

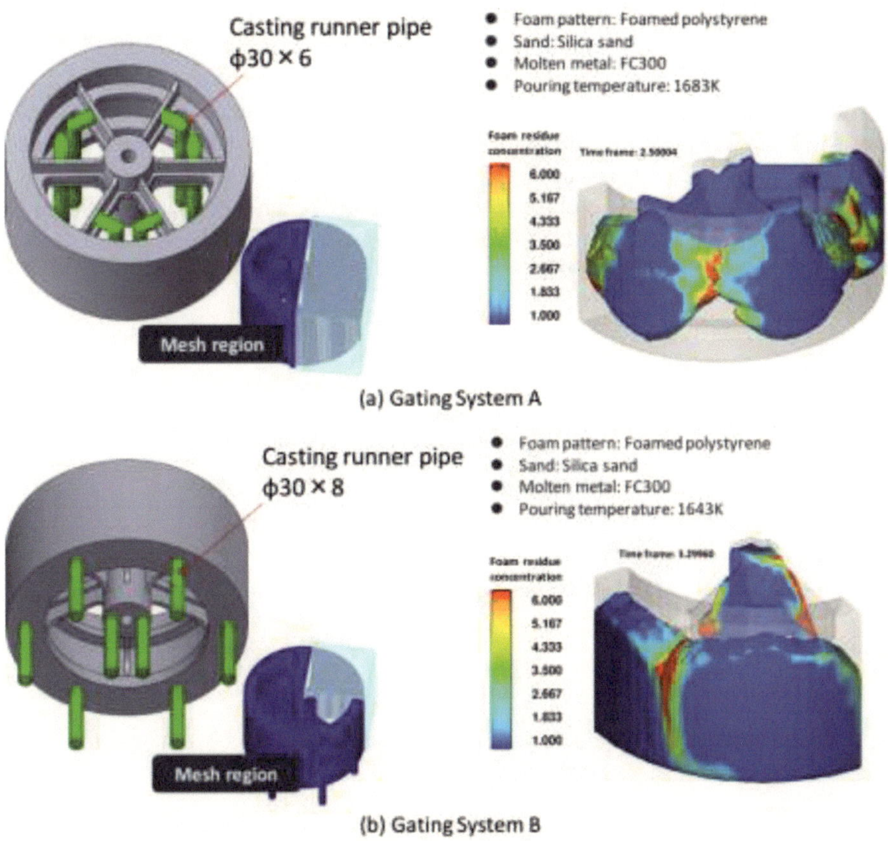

Figure 3.
Results of flow analysis for capstan drum.

pouring temperature is increased to prevent residue defects, the temperature at the completion of filling will also be high, and therefore, to prevent shrinkage defects the pouring temperature should not be increased. Therefore, to prevent both residue defects and the formation of a shrinkage cavity, a new casting plan that allows the suppression of residue defects without increasing the pouring temperature is required.

3.2 New gating system for full mold casting foam residue traps

By installing a gate at the bottom of the rim, the metal cannot avoid merging in the rim. If the pouring temperature cannot be increased, residue defects are more likely to occur. This requires the discharge of metal that contains residue from the rim. Therefore, we propose a new gating system: foam residue traps. For shrinkage nests, this system allows us to lower the temperature of the pouring metal. Although lowering the pouring temperature may increase the number of defects in the residue, a trap is installed on the side of the rim at the position where the metal joins with the metal to collect the residue, and the metal that is free from residue is recovered. This makes it possible to reduce residue defects.

In order to verify the effect of discharging molten metal entrapped by residue traps, a gating system was designed as shown in **Figure 4**. CFD analysis was performed for this. The analysis conditions are the same as in the previous section, except that a residue trap was installed.

The results of the CFD analysis are shown in **Figure 5**. The results of the analysis are velocity vectors of the metal, which are cross-sections based on the center of the height of the capstan drum. Horizontal cross-sections are set at −120 mm, 0 mm, and 120 mm height, respectively. At the 0 mm and 120 mm cross-sections, there is a vector of the flow from the inside of the residue trap to the rim. Therefore, metal that is entrapped in the residue may flow into the rim. Since the interface between the foam model and the metal is filled with molten metal with resistance due to combustion, it is considered that convection is formed by a flow that has no place to go. Therefore, it is necessary to design an appropriate casting plan to prevent the backflow of metal to the rim side due to convection inside the residue trap.

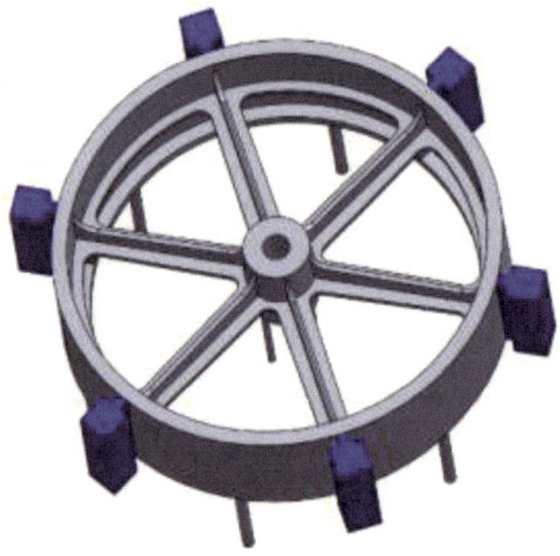

Figure 4.
Capstan drum with traps for preventing foam residues.

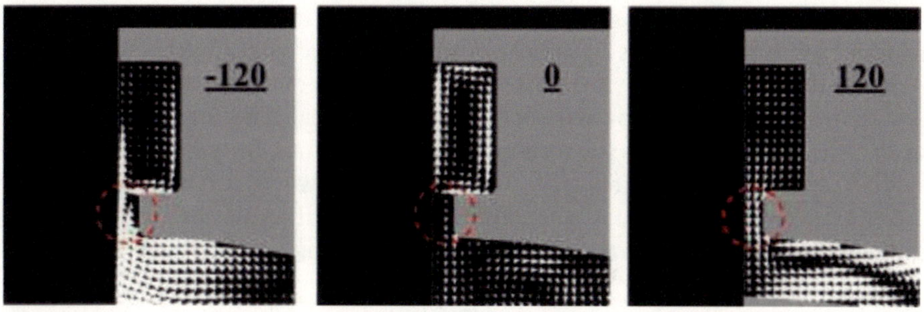

Figure 5.
The velocity vectors on the horizontal sections for the foam residue trap.

3.3 Optimal design of residue traps

Based on the results of the analysis in the previous section, the bottom of the rim is considered the best location for the gate, but in any case, residue defects will be generated by molten metal joining in the rim. In this study, a residue trap is proposed as a gating system for collecting molten metal that can join together and directly collect metal that is entrained with residue, and a new casting method for large castings is presented.

The optimal design of the residue trap is used for the capstan drum shown in **Figure 2**. In optimization, the dimensions defined as design variables are varied within a set range to find the optimal combination. This combination increases with the number of design variables, and the time required for optimization becomes enormous. Therefore, the number of design variables must be set to the minimum necessary.

The basic shape of the residue trap and the locations where the design variables were set are shown in **Figure 6**. The design variables were defined with five variables, x_1 to x_5. Six of these residue traps are installed evenly across the sides of the rim and are interlocked depending on the value of the design variables, all of which have the same shape.

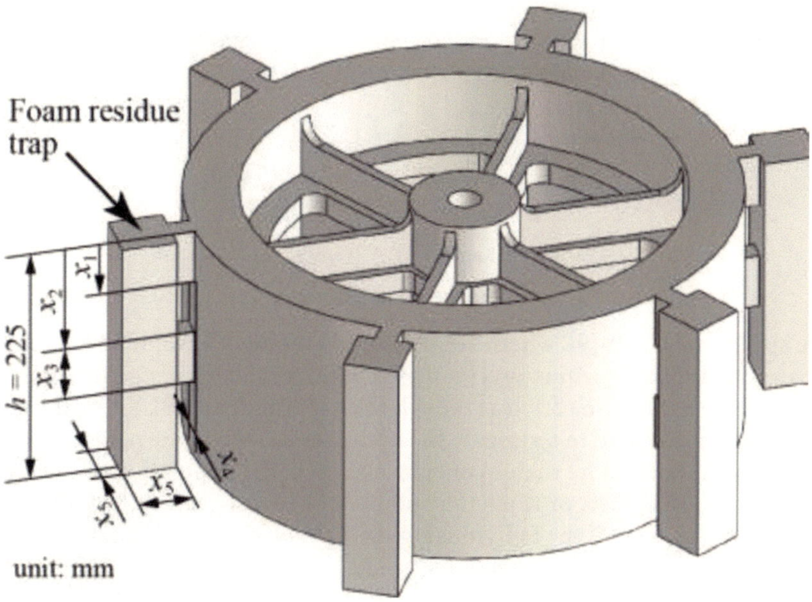

Figure 6.
Foam residue traps of capstan drum and design variables.

Three objective functions are defined as evaluation indices for optimization. The first objective function, J_1, is an indicator of the amount of residue in the rim. The residue in this CFD software is defined as the diffusion of the transport equation equal to the volume of the foam model replaced by metal. This amount of residue can actually be calculated as the residue volume fraction a_k for each calculated cell k, which represents the volume fraction of the residue contained in the cell in terms of solid time equivalent, and J_1 is defined as the following equation, where J_1 is this residue volume fraction averaged for each cell in the rim.

$$J_1 = \frac{\sum_{k=1}^{n} a_k}{n} \tag{4}$$

where n is the number of cells in the rim. In other words, J_1 represents the volume fraction of the residue in the entire rim. The smaller the value of J_1, the lower the residue defects are expected to be in the actual castings. In this capstan drum simulation, the number of cells in the rim is n = 32575, where each side is a cube of 5 mm.

The second objective function, J_2, is defined as the contact area between the rim and the residue trap. J_2 is calculated using the design variables x_1, x_3, and x_4 by the following equation.

$$J_2 = x_1 x_4 + x_3 x_4 \tag{5}$$

The smaller the value of J_2, the easier it is to remove the residue traps and the better the shape.

The third objective function, J_3, is defined as the volume of the residue trap itself, which is calculated using the design variable x_5 and the height of the residue trap, h, as follows.

$$J_3 = x_5^2 h \tag{6}$$

The smaller the value of J_3, the higher the material yield and the better the gating system.

Based on these definitions, we added geometric constraints and finally formulated this optimization problem.

$$\begin{gathered} Minimize\ J_1(x_1,,\ldots,,x_5), J_2(x_1,,x_3,,x_4), J_3(x_5) \\ \text{Subject to } 0\,mm \le x_1 \le x_2 \le 225\,mm \\ 0\,mm \le x_2 + x_3 \le 225\,mm \\ 10\,mm \le x_4 \le 30\,mm \\ 30\,mm \le x_5 \le 120\,mm \end{gathered} \tag{7}$$

We applied NSGA-II [9], a multi-objective genetic algorithm, to this optimization problem. The algorithm was set up with 50 individuals per generation, congestion tournament selection as the selection method, and BLX-α as the mating method. We also decided to terminate the optimization when the percentage of individuals generated more than 2 generations ago in the population of the focused generation exceeded 70%, or when the number of generations reached 30.

The distribution of all the individuals generated by the optimization for the objective function values is shown in **Figure 7**.

The total number of individuals generated was 775. This figure shows that the volume fraction of residue J_1 and the volume of residue trap J_3 are in a trade-off relationship with each other. C in the figure shows the objective function values of the residue traps in which no residue defects occurred in the preliminary experiments. Therefore, in this optimization we selected the individual that is superior for both objective functions as the optimal solution for this C. The final optimal shape is shown in **Figure 8**. The values of the design variables of the optimal solution are shown in **Table 1** and those of the objective function are shown in **Table 2**.

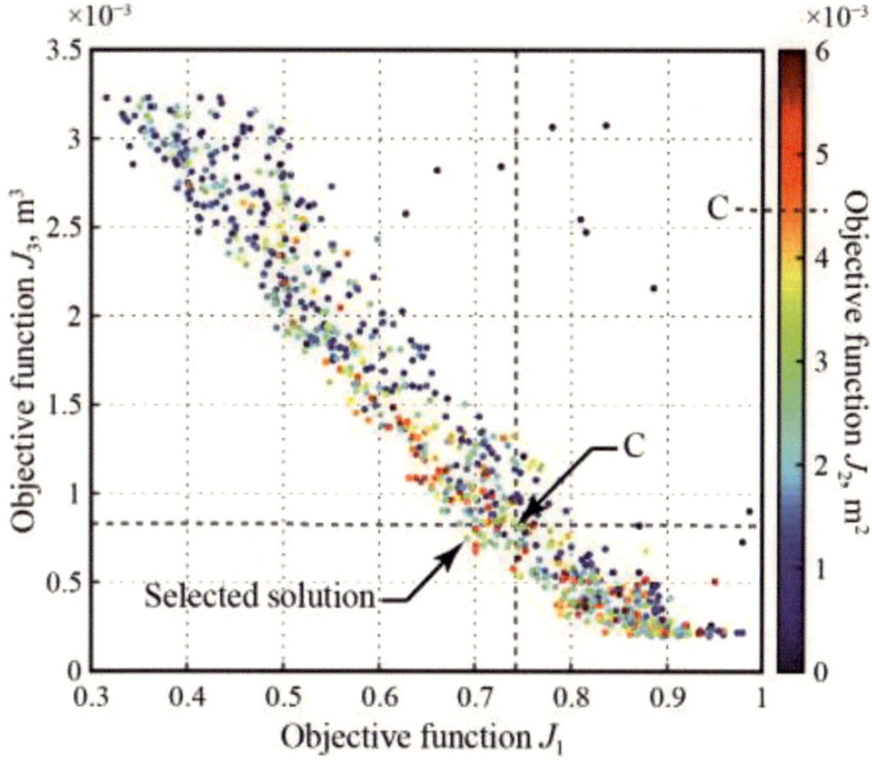

Figure 7.
Distribution of all analyzed individuals in optimization.

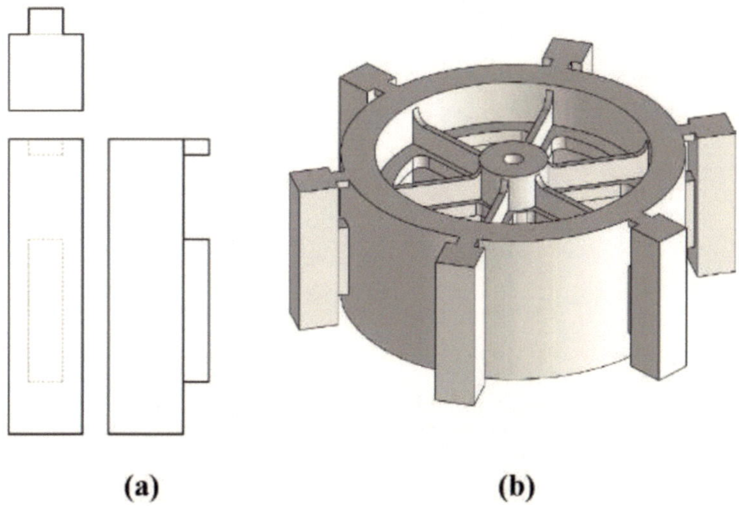

Figure 8.
Optimized shape of foam residue trap. (a) Foam residue trap. (b) Whole shape of casting.

x_1 [mm]	x_2 [mm]	x_3 [mm]	x_4 [mm]	x_5 [mm]
11.58	76.50	108.82	27.05	57.57

Table 1.
Values of design variables for optimal solution.

J_1 [−]	J_2 [m²]	J_3 [m³]
6.906×10^{-1}	3.257×10^{-3}	7.456×10^{-4}

Table 2.
Values of objective functions for optimal solution.

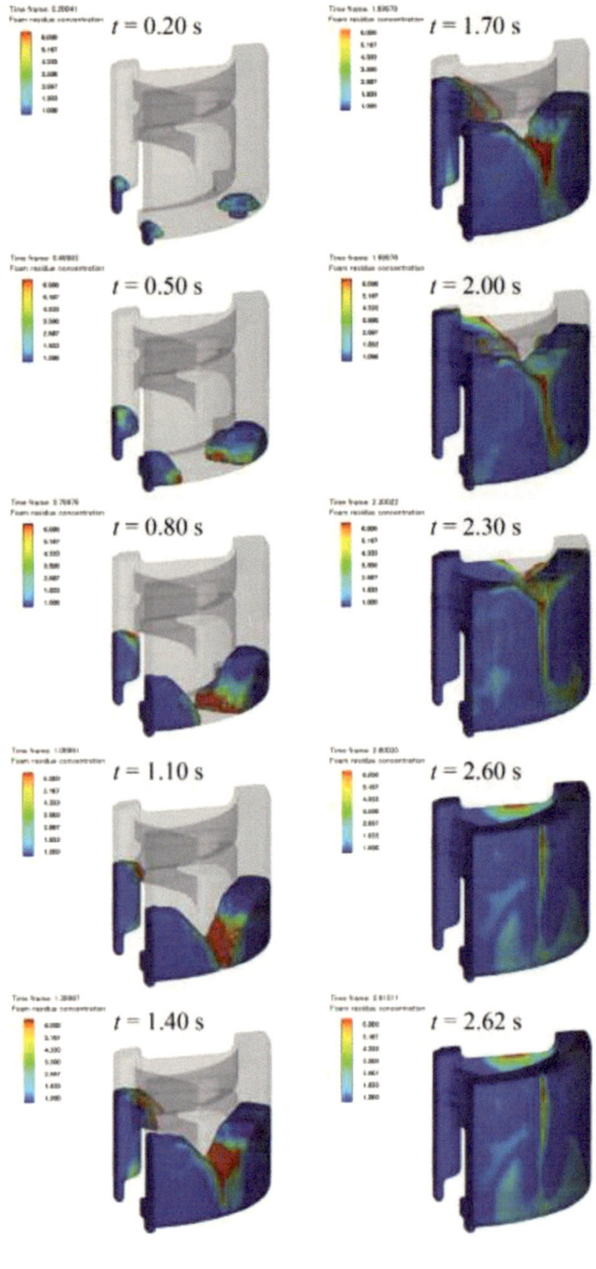

Figure 9.
Fluid behavior without foam residue trap.

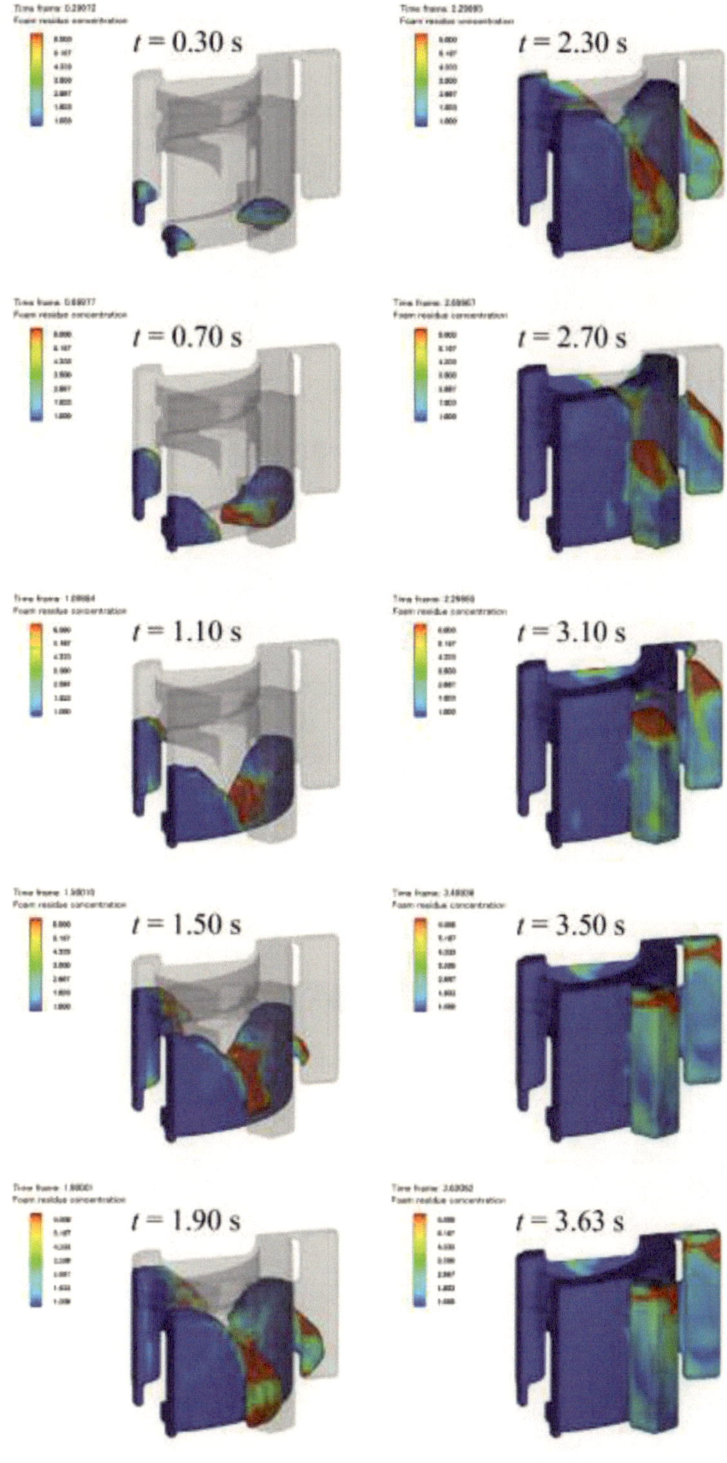

Figure 10.
Fluid behavior with shape-optimized foam residue trap.

In order to verify the effectiveness of the optimal shape, CFD analysis was performed and compared between the conventional method without a residue trap and the residue trap method with the optimal shape. The flows for the conventional and residue-trapping methods are shown in **Figures 9** and **10**, respectively.

In the conventional gating system, the metal flow shows that the residue is spreading inside the rim, and it is thought that the following factors influence the development of residue defects. The volume fraction of the residue in this case was $J_1 = 1.013$. On the other hand, the residue trapping method has an opening in the residue trap where the metal containing residue agglomerates, through which residue is effectively collected into the trap. The part of the rim that is not in contact with the residue trap suppresses the backflow of metal containing residue. Due to these effects, the optimal shape of the residue trap is efficient for collecting residue and effective for capturing residue. The volume fraction of residue is 0.6906, which is more than 30% lower than that of the conventional method.

4. Verification of effectiveness by casting experiments

Casting experiments were conducted to verify the effectiveness of the optimally designed residue trap.

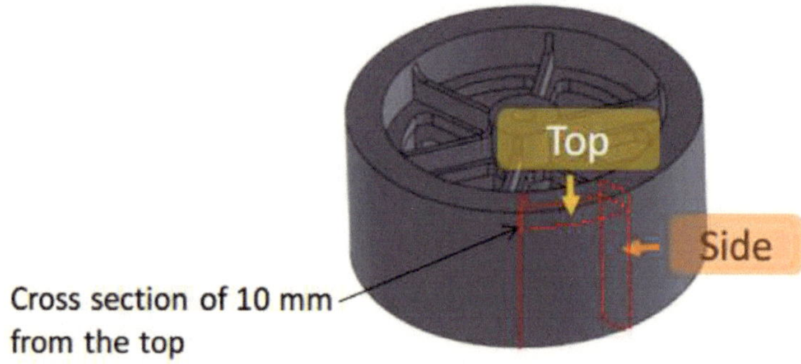

Figure 11.
Cross section of capstan drum for experiments.

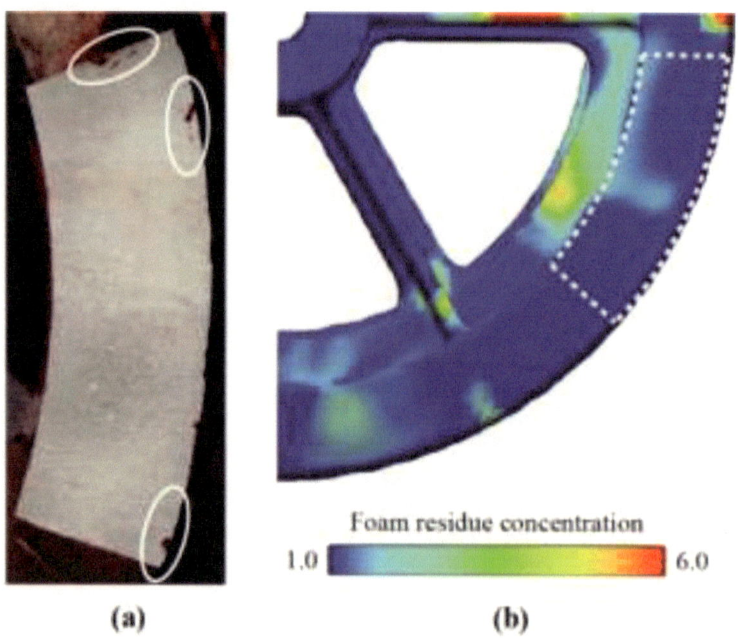

Figure 12.
Evaluation of foam residue defect for casting without foam residue trap. (a) Experimental result. (b) Flow analysis result.

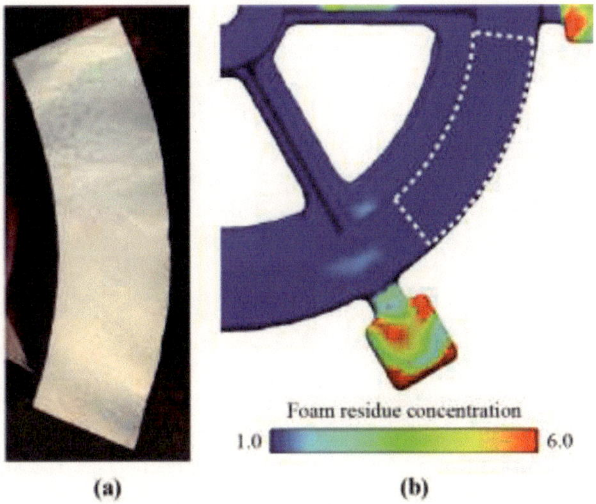

Figure 13.
Evaluation of foam residue defect for casting with shape-optimized foam residue trap. (a) Experimental result. (b) Flow analysis.

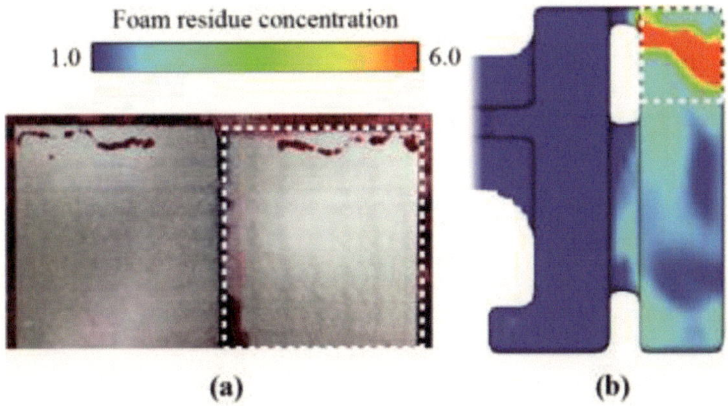

Figure 14.
Evaluation of foam residue defect in foam residue trap. (a) Experimental result. (b) Flow analysis result.

In this study, two types of capstan drums were cast: one with a conventional method without a residue trap and the other with an optimized residue trap method. The temperature of the metal was set to 1623 K, and the other conditions were the same as in the experiments in Section 2. After casting and cooling, the product was cut to make a test specimen, and the cut surface was color-checked to compare the defects. As shown in **Figure 11**, the position where the product part was cut was set to a horizontal plane 10 mm downward from the top of the rim because the residue tends to accumulate on the top of the rim.

The experimental results of the conventional plan and the corresponding simulation results are shown in **Figure 12**. This result suggests that this was caused by the residue of the foam model generated during the filling of the molten metal and remaining in the rim of the drum. The simulation results in **Figure 12(b)** indicate that residue defects are more likely to occur around the specimen, although the simulation does not fully predict the distribution of residue.

Figure 13 shows the experimental results of a residue trapping method using the optimal shape and the corresponding simulation results.

Slight shrinkage nests were observed in the specimens, but no residue defects were observed. Similarly, in the simulation results, there was almost no residue around the specimen. Therefore, the residue traps were also cut in the same way as the test specimens, and the color inside each trap was checked. The results and the corresponding simulation results are shown in **Figure 14**. Residue defects were observed in the residue trap, indicating that the residue generated inside the product was captured well by the trap. Therefore, this study confirms that the residue trapping method using the optimal shape effectively reduces residue defects.

5. Conclusion

In this study, a new residue trapping method is proposed to suppress defective residues, which are a problem in manufacturing large castings by the full mold casting method. To maximize the performance of the residue trap, the shape of the trap was optimized by using CFD simulation. The effectiveness of the optimized residue traps was verified by casting experiments, which showed that the effectiveness of the optimized traps can reduce residue defects better than those without traps.

Author details

Yuto Takagi, Masahiro Inagaki and Ken'ichi Yano*
Mie University, Mie, Japan

*Address all correspondence to: yanolab@robot.mach.mie-u.ac.jp

References

[1] J. Thiel and S. Ravi: Analysis and Prediction of Casting Defects Using Casting Simulation. Proc. 71st World Foundry Congress, Bilbao (WFO) (2014)

[2] K. Kanazawa, K. Yano, J. Ogura, and S. Baba: New Curve Optimization Method and Its Application to Shape Design for Die Casting Using a CFD Simulation. No. 62541 in Proc. ASME 2013 IMECE (2013)

[3] T. Maruyama, N. Goto and T. Kobayashi: Thermal decomposition products of expanded polystyrene pattern in evaporative pattern casting of cast iron. Journal of Japan Foundry Engineering Society. 2009;81(3):117-122

[4] S. Koroyasu and M. Matsuda: Effects of high insulating coat on cooling rate of aluminum alloy in evaporative pattern casting process. Journal of Japan Foundry Engineering Society. 2004;76(8):678-686

[5] M. Karimian, A. Ourdjini, M. Hasbullah and H. Jafari: Effect of pattern coating thick-ness on characteristics of lost foam Al-Si-Cu alloy casting. Transactions of Nonferrous Metals Society of China. 2012;22(9):2092-2097

[6] C. W. Hirt and M. R. Barkhudarov: Modeling the Lost Foam Process with Defect Prediction. Flow Science Tech. Note. 1997;FSI-97-TN45

[7] C. W. Hirt and M. R. Barkhudarov: Predicting Defects in Lost Foam Castings. Modern Casting (2002)

[8] C. W. Hirt: Modeling the Lost Foam Process with Defect Predictions Progress Report: Lost Foam Model Extensions, Wicking. Flow Science Tech. Note. 1999;FSI-03-TN45-1

[9] K. Deb, A. Pratap, S. Agarwal and T. Meyarivan: A Fast and Elitist Multiobjective Genetic Algorithm: NSGA-II. IEEE Transactions on Evolutionary Computation. 2002;6(2):182-196

Chapter 6

The Physical Chemistry of Steel Deoxidation and Nozzle Clogging in Continuous Casting

María-Guadalupe González Solórzano, Rodolfo Morales-Dávila, Jafeth Rodríguez Ávila, Carlos Rodrigo Muñiz-Valdés and Alfonso Nájera Bastida

Abstract

Nozzle clogging in continuous casting of steel originates by the adherence of alumina particles and other oxides, precipitated during the liquid steel deoxidation, on the refractory material's surface. Hence, these particles' nucleation and growth rates in supersaturated melts are analyzed considering, specifically, the role of the interfacial tensions between alumina, silica, and other oxides and the liquid metal. Weak steel deoxidizers like silicon do not need high supersaturations favoring high nucleation rates, giving particles' narrow size distributions thanks to fast diffusion and Ostwald-ripening coagulation. Strong deoxidizers, like aluminum, need high supersaturation levels leading to broad size distributions. Besides, the morphology of these particles depends on the nucleation and growth mechanisms. The adhesion forces among the deoxidation particles, forming clusters, depending on the morphology and the oxide's chemistry. The stability of the nozzle's clog, adhered to the nozzle's wall, depends on the interface tensions between the melt and the nozzle's refractory surface and between the melt and the inclusion. The results obtained here help set up basic recommendations in steel refining and materials specifications of casting nozzles.

Keywords: steel deoxidation, nozzle clogging, melt supersaturation, nucleation, growth, interfacial tension

1. Introduction

Clogging of mold nozzles, also called submerged entry nozzles (SEN), disrupts the steel casting process, affecting the caster's productivity. The nozzle clogging produces inconsistent flow and temperature variations, steel level fluctuations in the mold, impairment of steel quality, and the steel casting's abrupt interruption. Clogging starts when solid compounds, mainly steel skull and non-metallic inclusions, are non-uniformly deposited at the inner nozzle wall, at some typical preferential zones characterized for neighboring dead flow conditions [1–5]. These inclusions have as primary sources: (1) The reaction between the dissolved oxygen with the deoxidizers [6–9]; (2) re-oxidation in the tundish or the nozzle [10, 11]; and (3) the entrainment of slag or refractory particles [11–14]. Researchers who have worked on the determination of inclusions sources and clogging recognize that

 IntechOpen

the deposited inclusions at the nozzle wall are mainly alumina inclusions [7–9, 15, 16]. Steel re-oxidation occurs due to possible air aspiration under the flow control valves (slide gate) to maintain the entry flow to the molds [17–19]. Besides, regardless of the refractory nozzle composition (alumina-graphite, zirconia, and magnesia), the steel melt infiltrates the refractory and removes the protective surface [11, 20], allowing the entrapment of refractory particles and inclusion attachment at the nozzle wall.

The non-metallic inclusions come from the steelmaking process; several researchers have focused on studying the variables that induce the inclusion deposition at the inner nozzle wall producing the clogging phenomena [1, 11–14, 16, 21–22]. Steel chemistry and, in particular, steel grades containing titanium, like Ti-SULC, (Ti Stabilized Ultra-Low Carbon Steels), steels enhance the nozzle clogging due to the surface tension properties of this element in liquid steel [23–25]. The wetting of inclusions, rich in Ti oxide, assists in the clustering and compaction of particles. When the ratio Ti/Al is above a threshold, dictated by thermodynamics, the wettability of complex oxides of Ti and Al works intensifying the nozzle's clogging under the presence of oxygen. This series of papers provides an insight into the clogging phenomena while casting these steel grades.

In the present chapter, the authors deal with the physical–chemical aspects of the nozzle-clogging problem by inclusions originated through the deoxidation reactions of steel. Hence, to understand the fundamentals of the problem first, these particle's nucleation and growth are considered using non-equilibrium thermodynamics principles. Second, the influence of the steel refining processes on the deoxidation particle morphology and the relation with their further growth through aggregation and clustering mechanisms is under scrutiny. The analysis continues with studying the thermodynamics related to the oxide particle's adherence to the refractory. Finally, a dynamic analysis lets the establishment of balance among drag, buoyancy, adherence, and lift forces leading to a detachment criterion for an inclusion forming part of a first layer of the clogging. After this work, conclusions and recommendations are provided.

2. Non-equilibrium thermodynamics

2.1 Nucleation and growth rates of oxide inclusions

The nucleation of a foreign phase in an originally homogeneous solution, in the present case an oxide particle in a liquid solution of iron, is driven by the free energy of the reaction.

$$x[M] + y[O\} = M_xO_y \tag{1}$$

$$\Delta G_m = \Delta G_m^0 + RTln\frac{a_{M_xO_y}}{h_M^x h_O^y} = \Delta G_m^0 + RTlnQ \tag{2}$$

where Q is the activity quotient, in the thermodynamic equilibrium $\Delta G_m = 0$ and the Eq. (2) changes to,

$$\Delta G_m^0 = -RTlnK_E \tag{3}$$

where K_E is the equilibrium constant, which depends only on the temperature. Combining Eqs. (2) and (3), dividing the result between the molar volume of the oxide results in,

$$\Delta G_V = RTln \frac{Q}{K_E} = -RTln \frac{S_0}{V_0} \tag{4}$$

and S_0 is the supersaturation ratio given by,

$$S_0 = \frac{Q}{K_E} = \frac{K_{MO}}{K_E} = \frac{\left[h_M^x h_O^y\right]}{\left[h_M^x h_O^y\right]_E} = \frac{[\%M]^x [\%O]^y}{\{[\%M]^x [\%O]^y\}_E} \tag{5}$$

K_{MO} is the solubility product of the oxide. The free energy balance for the nucleation of an oxide precipitated in the homogeneous melt includes the volumetric free energy and the surface energy opposing to the stabilization and growth of the nucleus,

$$\Delta G = 4\pi r^2 \sigma_{PL} + \left(\frac{4}{3}\right)\pi r^3 \Delta G_V \tag{6}$$

Making zero the derivative of the free energy (**Table 1** [26, 27] reports the molar Gibbs free energy of deoxidation reactions of iron melts) allows the calculation of the critical radius for the onwards growth of the nucleus, obtaining,

$$r_c = \frac{-2\sigma_{PL}}{\Delta G_V} = \frac{2\sigma_{PL} V_O}{RTlnS_0} \tag{7}$$

The free energy required for the nucleation is obtained by substituting this radius in Eq. (6),

$$\Delta G = \frac{16\pi\sigma_{PL}^3}{3\Delta G_V^2} = \frac{16\pi\sigma_{PL}^3 V_O^2}{3(RTlnS_0)^2} \tag{8}$$

The nucleation rate is [26, 28],

$$I = exp\left[\frac{16\pi\sigma_{PL}^3 V_O^2}{3k_B R^2 T^3 (\ln S_O)^2}\right] exp\left(\frac{-\Delta G}{k_B T}\right) = Aexp\left[\frac{-16\pi\sigma_{PL}^3 V_O^2}{3k_B R^2 T^3 (\ln S_O)^2}\right] \tag{9}$$

$$A = exp\left[\frac{16\pi\sigma_{PL}^3 V_O^2}{3k_B R^2 T^3 (\ln S_O)^2}\right] \tag{10}$$

where k_B = 1.38X10^{-23} J K^{-1}, is the Boltzmann constant, A = 10^{32} m^{-3} s^{-1} is the frequency factor, V_O is the oxide's molar volume, σ_{PL} is the interfacial tension

Reaction	log $K_{eq.}$ (at 1873 K)
Al_2O_3 (s) = 2 Al + 3 O	−55.76 (= 11.80–62790/T)
ZrO_2 (s) = Zr + 2 O	−36.24 (= 21.76–57000/T)
MgO (s) = Mg + O	−32.85 (= 12.45–38050/T)
SiO_2 (s) = Si + 2 O	−19.39 (= 11.58–30400/T)
MnO (s) = Mn + O	−5.55 (= 6.70–15050/T)
CaO (s) = Ca + O	−42.76 (= 7.77–33700/T)

Table 1.
Equilibrium constants [26, 27].

between liquid iron and an oxide particle, R is the gas constant, and T is temperature. The critical supersaturation S_O^* is the minimum one to nucleate one nucleus m^{-3} s^1. Hence, making I = 1 m^{-3} s^{-1} in Eq. (9) implies $S_O^* = S_O$, and gives,

$$S_O = exp\left[\frac{V_O}{RT}\sqrt{\frac{16\pi\sigma_{PL}^3}{3k_B TlnA}}\right] \tag{11}$$

The supersaturation of iron melts in contact with different oxides was measured through electrochemical methods [29]. The critical supersaturation, S_O^*, depends on the oxide's nature and surface tension, as seen in the precedent equation. The experimental results provide the inclusions sizes' statistical dispersion directly in the sense that the larger surface tensions, contributing to the opposing free surface energy to the nucleation, lead to larger size dispersions [26]. **Table 2** shows the experimental results of supersaturation experiments reported in Ref. [26] using different metal deoxidizers. The second and third columns are the molar volume and the interfacial tension between the oxide and the melt, the fourth and fifth, the experimental and calculated critical supersaturations through Eq. (11). The sixth and seventh columns include the experimental and corrected supersaturations, respectively. Revision of these data shows that the magnitudes of the calculated critical supersaturations are larger than the experimental critical supersaturations. The other observations are the small magnitudes of the experimental supersaturation. The large differences behind the critical supersaturations are due to the experimental data's nature as they correspond to the metal bulk property. On the other hand, the other magnitude (the corrected supersaturations) is theoretically related to the nuclei's curvature, as is schematized in **Figure 1**. The curvature raises the gradients of concentration of all solutes, M (Al, Si, Mn, Mg, Ti, Zr, Ca) and O in the nuclei's tip and, consequently, the critical supersaturations are larger in the proximities of the tip than in the metal bulk. The curvature, given by r_C^{-1} is calculated through Eq. (7) and matching its magnitude by considering an embryo formed by two or three pairs of M-O dimers, (consulting for that purpose the atomic radius of the involved elements of the corresponding oxides in the Periodic Table).

The correction of the experimental supersaturations is possible through the Gibbs–Thomson's Equation which gives the ratio between both types of supersaturations as,

$$[S_{O,corr}]_{r=r} = [S_{O,exp}]_{r=\infty} exp\left[\frac{V_O}{RT}\frac{2\sigma}{r_c}\right] \tag{12}$$

OXIDE	Vo [m^3. Mol^{-1}]	σ_{PL} [J. m^2]	$S^*_{O\ (exp.)}$	S^*_O (cal.)	S_O	$S^*_{O\ (exp.)}$ corr	S_O corr.
MgO	1.10E-05	1.8	8.4	280	3.6 ~ 7.2e4	280.224	1801440
ZrO_2	1.01E-05	1.63	4.5	85	48 ~ 68	84.825	1093.3
Al_2O_3	8.60E-06	2.11	14.7	529	37 ~ 60	530.229	1803.5
CaO	1.65E-05	1.17	3.4	83			
CaO-Al_2O_3	1.47E-05	1.3	3.6	92	8.5E+04	90.936	2147100
SiO_2	1.13E-05	1.24	1	27			
MnO-SiO_2	1.36E-05	1	1.7	18	1.5	18.598	16.41

Table 2.
Experimental, calculated and corrected critical supersaturation degree for precipitation of oxides [26].

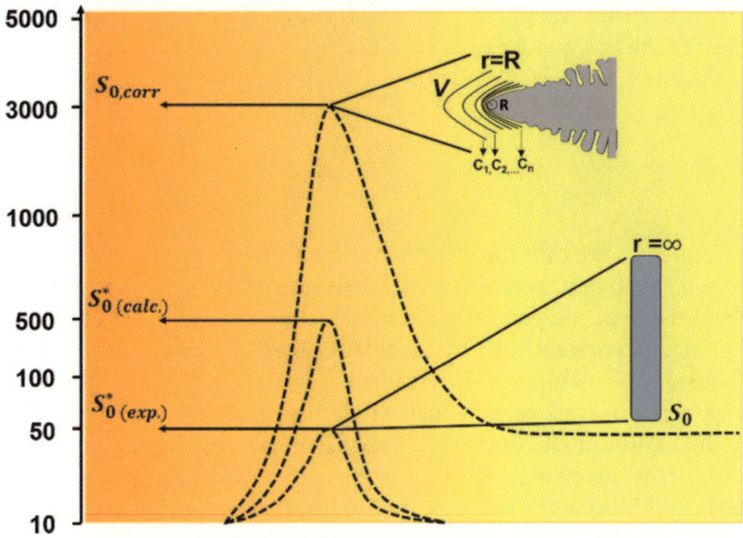

Figure 1.
Supersaturations of alumina close and far away from surfaces with finite and infinite curvature radius.

The nucleation process consumes short times of the order of microseconds [26, 29, 30], as shown schematically by **Figure 2.** The fluctuations of concentrations in the liquid structure, after the de-oxidant addition, require a critical supersaturation reached in the point (a) of this figure. Once reached the critical supersaturation, the embryo starts with a group of M-O dimers that develop into metastable structures becoming into a nuclei, and as the thermodynamic and kinetic conditions permit it, evolves in alumina with time, as seen in **Figure 3**. After stabilizing the nuclei, the local supersaturation decreases by diffusion process [31], due to the local consumption of M and O, to the point (b) in **Figure 2** where the nucleation kinetics overlaps with the diffusion process. The supersaturation continues decreasing and the

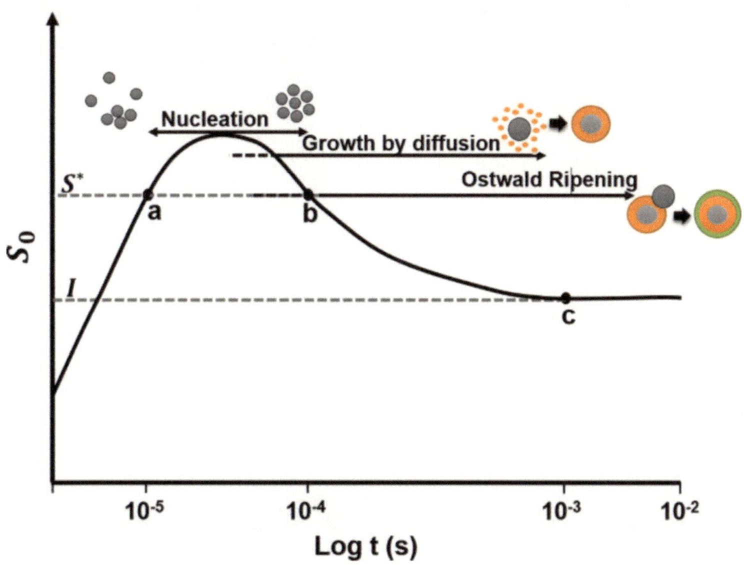

Figure 2.
Evolution of supersaturations with time during nucleation and initial growths by diffusion and Ostwald ripening mechanisms.

diffusion overlaps with the particle growth through the Ostwald ripening process summarized by the following expression [32],

$$\acute{r}^3 - \acute{r}_0^3 = \alpha k_d t \tag{13}$$

$$k_d = \frac{2\sigma D_O V_O C_O}{RT(C_P - C_O)} \tag{14}$$

where C_O and C_P are the concentrations of dissolved oxygen and oxygen content in the oxide, respectively, and D_O is the diffusion coefficient of oxygen. Further growth phenomena of the particles, in industrial vessels, includes Stokes collisions, collisions among particles driven by turbulent flows deriving in aggregates of particles and clusters. Grown particles are easily floated out thanks to the bottom stirring with argon of steel ladles which carries these particles through their contact with gas bubbles, and melt convection making them contact the slag facilitating their absorption in this phase.

Figure 4a and **b** show the effect of the supersaturation on the nucleation kinetics on the nucleation rates of different oxides recalculated from reference [26]. There are two important differences, the first is that the nucleation rates are considerably lower than those reported in Ref. [26] and the second is the larger supersaturations required to precipitate the MgO shown in **Figure 4b**. Another important feature is the small supersaturations required for the precipitation of silica and manganese silicate. Therefore, those particles requiring small or relatively small supersaturations yield the largest nucleation rates meaning, physically, the fast generation of million of nuclei distributed inside the reaction and diffusion boundaries.

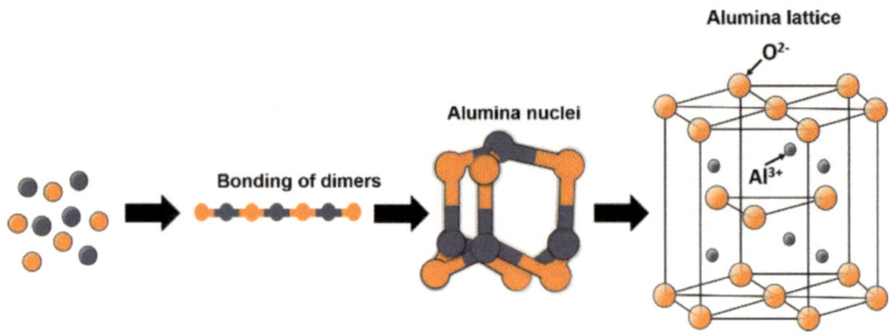

Figure 3.
Nucleation of an alumina lattice from the union of Al-O dimers.

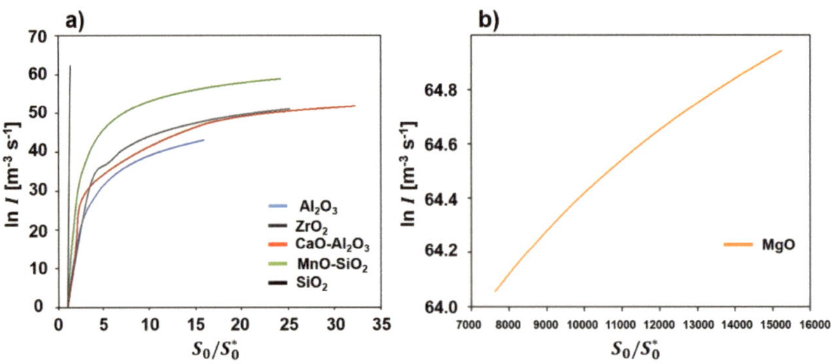

Figure 4.
Effect of supersaturation ratios on nucleation rate at 1600°C, (a) various metal deoxidizers, (b) with magnesium.

The main obstacle of a particle to born is the creation of a new surface preceding the formation of a small volume in a hosting matrix, as is seen in **Figure 3**, with a different structure. Those particles requiring larger supersaturations would, eventually, nucleate smaller populations of particles with a broader size distribution once their first step of development ends. The precipitates characterized by high and low supersaturations are schematized in **Figure 5a** and **b**, respectively, showing the period for each one of them. The first case will yield numerous small particles with limited growth as the supersaturation decays providing a narrow size distribution. The longer times lead to larger diffusion and growth time scales yielding broader size distributions in the second case. The time scales change to exponentially larger ones when the precipitates of solid-state transformations take place. For example, steel aging by nitrogen diffusion consumes long times as the diffusion coefficients of this interstitial element are very small in ferrite or austenite phases at 300°C [32, 33] compared with its diffusion coefficient at 1600°C [34], as is schematized in **Figure 5c** and **d.** Hence, the interfacial tension between the melt and the nuclei governs the nucleation rate, as seen in **Figure 6**. Alumina has one of the lowest nucleation rates as its interfacial tension is large, and its crystals will have a broad spectrum of sizes after the nucleation and the diffusion end.

2.2 Alumina morphology

Alumina may acquire a wide diversity of morphologies depending on the concentrations of oxygen and the deoxidant. Accordingly, the initial supersaturation ratio influences the morphology of alumina. **Figure 7** shows a scheme of the relation between oxygen concentrations and the deoxidant with the particle morphology [35]. In oxidized melts, the inclusions are rounded spheroids, as the oxygen activity decreases the surface roughness develops reaching the stage of dendritic precipitation. With further deoxidation the morphology changes to faceted, disks and crystalline rhomboids.

The heterogeneous nucleation of alumina particle also yields characteristic morphologies. Other foreign particles catalyze alumina's nucleation in the melt, such as those of metastable iron oxide precipitated in the steelmaking furnace and manganese silicates in the refining ladle [36]. A catalyzed nucleation process means that

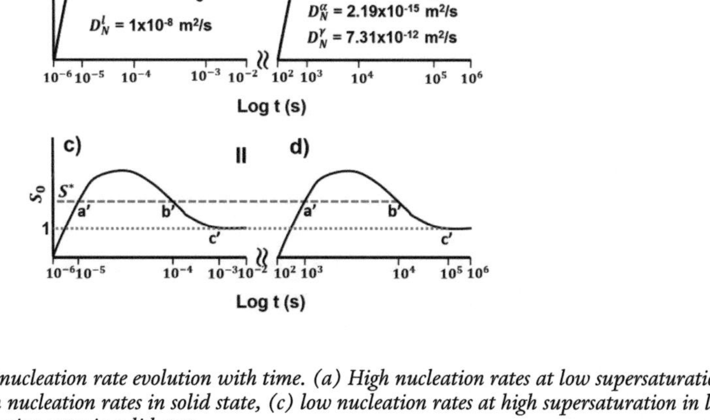

Figure 5.
Schematics of nucleation rate evolution with time. (a) High nucleation rates at low supersaturation in liquid state, (b) high nucleation rates in solid state, (c) low nucleation rates at high supersaturation in liquid state, (d) low nucleation rates in solid state.

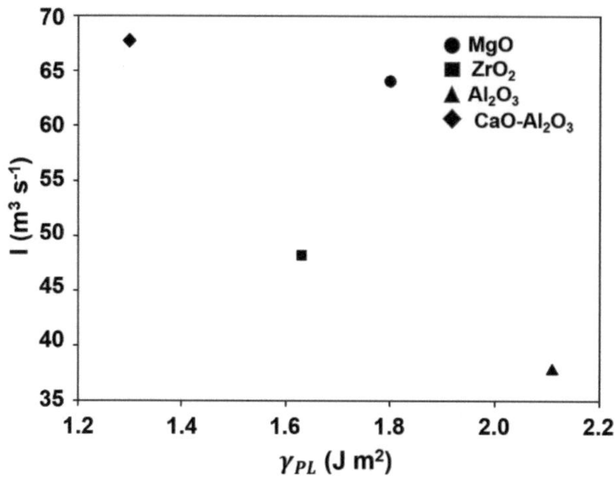

Figure 6.
Nucleation rate as a function of surface tension between the particle and the melt at 1600°C.

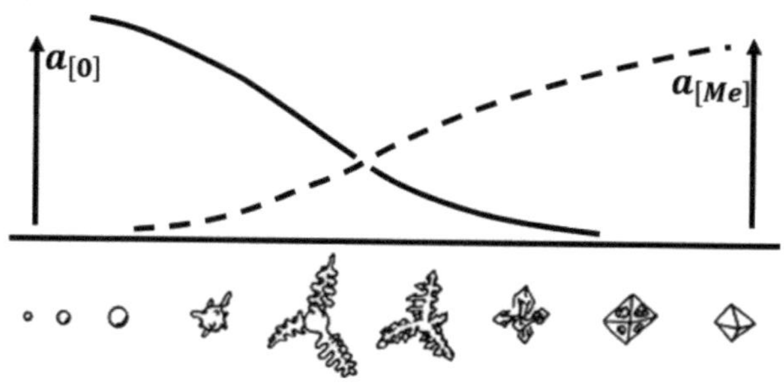

Figure 7.
Evolution of the growth shapes of oxide inclusions as a function of the local oxygen activity (solid line) and deoxidizer activity (dashed line) according to Steinmetz [35].

the foreign particle decreases the supersaturation required by the precipitation process of the particle through a decrease of the free energy according to [32],

$$\Delta G_{het} = \left\{ \frac{-4}{3} \pi r^3 \Delta G_v + 4\pi r^2 \sigma_{PL} \right\} S(\theta) \tag{15}$$

where

$$S(\theta) = \frac{(2 + cos\theta)(1 - cos\theta)^2}{4} \tag{16}$$

Note that except for factor S(θ) this expression is the same as that obtained for homogeneous nucleation, Eq. (6). S(θ) has a numerical value ≤1 dependent only on θ, i.e. contact angle and the nucleus's shape. A catalyzed alumina particle will have a morphology depending on the oxide's nature over which it nucleates as shown in **Figure 8** [36]. If the foreign o catalyzer particle is a silicate, already existent given its high nucleation rate, a chemical reaction intervenes,

$$2[Al] + (MnO{\bullet}SiO_2) = (Al_2O_3) + [Mn] + [Si] \tag{17}$$

if the particle is iron oxide, then

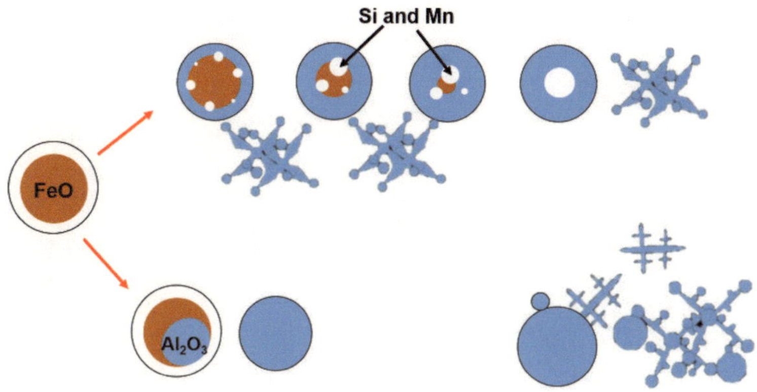

Figure 8.
Heterogeneous nucleation of alumina on iron oxide and silicate particles [36].

$$2[Al] + 3(FeO) = (Al_2O_3) + 3[Fe] \tag{18}$$

The interfacial tension between the alumina and silicate particles is reduced due to the chemical reaction, a layer of alumina, (it is initially nucleated and the alumina layer grows by the reaction), surrounds the particle of silicate making sluggish the diffusion of aluminum through the alumina layer to continue the reaction. During the process, some particles of Mn and Si precipitate as products of the reaction (17). No dendrites were observed and there is the presence of needle like clusters with disk type terminations indicating a growth under small supersaturations, once the oxygen content decreases. Iron oxide suffers a rapid reduction by aluminum, reaction (18) and the product is an alumina particle without other phases. Other alumina particles nucleate heterogeneously on the original alumina particles and yield dendritic morphologies.

2.3 Further growth of inclusions

After concluding the nucleation and growth by diffusion and Ostwald ripening, the inclusions continue their growth through direct collision. In a Stokes flow regime, the probability for collision among inclusions of sizes R_1 and R_2 is [37],

$$w_s = \frac{2}{9} g \frac{\Delta\rho}{\mu} |R_1 - R_2| (R_1 + R_2)^3 \tag{19}$$

and the probability of collision under the action of turbulent eddies is expressed by the Saffman's Equation as [38],

$$w = 1.3\sqrt{\pi}(R_1 + R_2)^3 \sqrt{\varepsilon/\nu} = 7.2|R_1 - R_2|(R_1 + R_2)^3 \{Noexp(-\alpha R_1) Noexp(-\alpha R_2)\}/2 \tag{20}$$

In Stokes regime, the highest collision probability is observed when the inclusions have large size differences as seen in **Figures 9** and **10** indicates that collisions of silicate inclusions smaller than ten μm, have low probabilities for collisions under turbulent flow conditions and those with large dimensions have higher collision probabilities. **Figure 11** shows the corresponding collision probabilities for alumina inclusions and particles as small as three μm yield the highest probabilities for a collision. Therefore, alumina inclusions grow from microscopic inclusions to large aggregates and clusters by collisions among small particles forming aggregates [40].

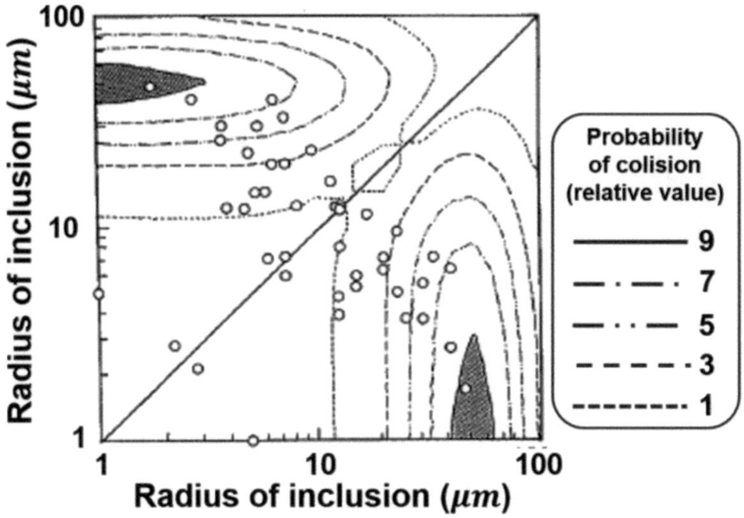

Figure 9.
Comparison of the collision probability (Stoke's model) with the radius of the inclusions. Miki Y, Kitaoka H, Sakuraya T, Fujii T. mechanism for separating inclusions from molten steel stirred with a rotating electro-magnetic field. ISIJ Int. 1992;32:142–149. DOI: 10.2355/isijinternational.32.142. Reproduced with permission [39].

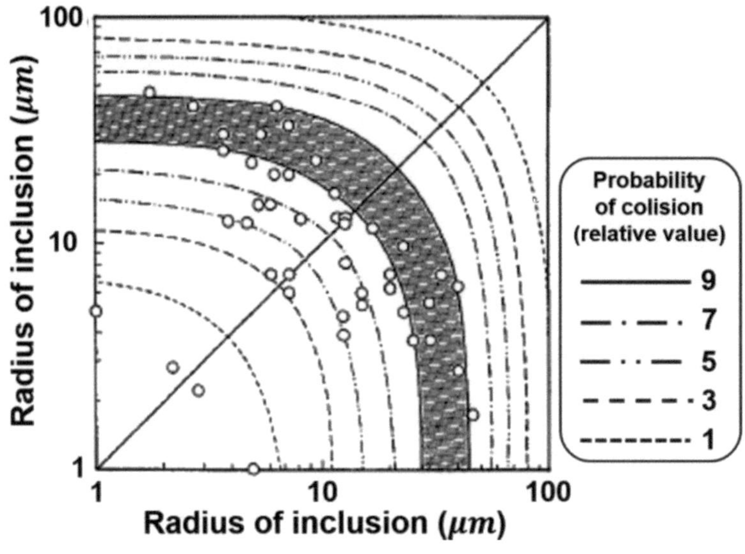

Figure 10.
Comparison of the collision probability (Saffman's model) with the radius of the inclusions. Miki Y, Kitaoka H, Sakuraya T, Fujii T. mechanism for separating inclusions from molten steel stirred with a rotating electro-magnetic field. ISIJ Int. 1992;32:142–149. DOI: 10.2355/isijinternational.32.142. Reproduced with permission [39].

Roughly speaking, in the turbulence regions, in a bottom stirred ladle, the ratio between Stokes and Safmann's regimes is approximately $6X10^4$. Thereby, deoxidation during steel tapping in bottom stirred ladles is the most indicated step to deoxidize and grow inclusions by turbulent collisions. This is particularly applicable to the growth of alumina particles. Indeed, vigorously stirred melts at tapping times, lead to cleaner steel heats [41]. In the secondary refining of steel, where the turbulence levels are considerably smaller, this ratio decreases and is the highest in the argon-plume regions, while in the top layer of the bath, the Stokes regime dominates.

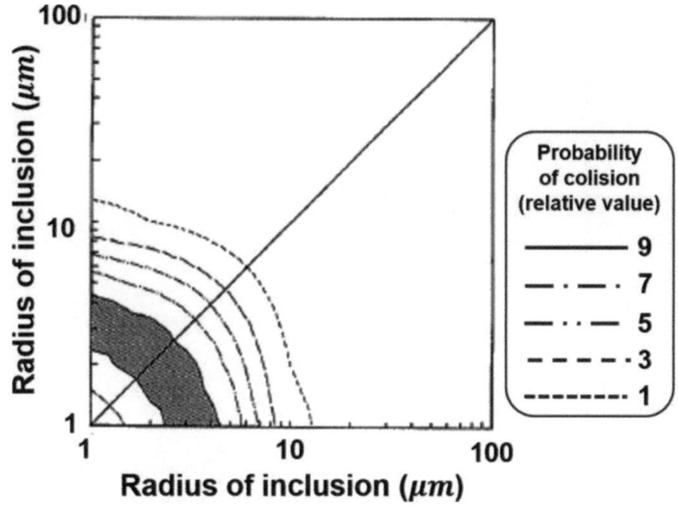

Figure 11.
Prediction of the collision probability (Saffman's model) for alumina inclusions. Miki Y, Kitaoka H, Sakuraya T, Fujii T. mechanism for separating inclusions from molten steel stirred with a rotating electro-magnetic field. ISIJ Int. 1992;32:142–149. DOI: 10.2355/isijinternational.32.142. Reproduced with permission [39].

2.4 Bond strength among particles

Another important aspect of inclusions growth is the stability of aggregates and clusters, forming large particles that float out of the bath faster as larger are their sizes. Strong bond strengths are desirable as once the particles form aggregates or clusters, their integrity must prevail, avoiding the generation of smaller particles, by breaking processes due to turbulence, which may bring on floatation slowness. The variety of bonds among inclusions with a wide spectrum of morphologies is simplified into three basic cases, sphere-sphere, sphere-plate and plate-plate, see **Figure 12**. According to thermodynamic calculations of surface tensions, the plate-plate geometry yields the largest bond strength, as suggested by **Figure 13** [42]. The bond strength, eventually, will increase by the thickenings of the neck formed by the union of two particles of alumina through diffusion processes according to,

$$\frac{x^5}{R^2} = \frac{K_1 \sigma_{PP} V_0}{RT} D_V t \tag{21}$$

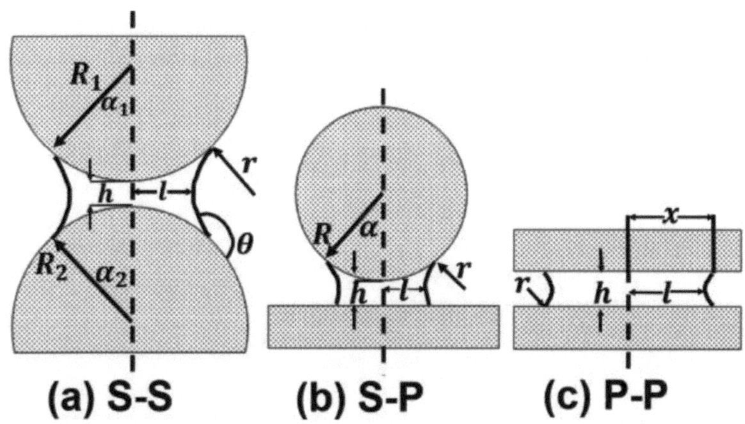

Figure 12.
Geometries of the gas cavity of different contact types [42].

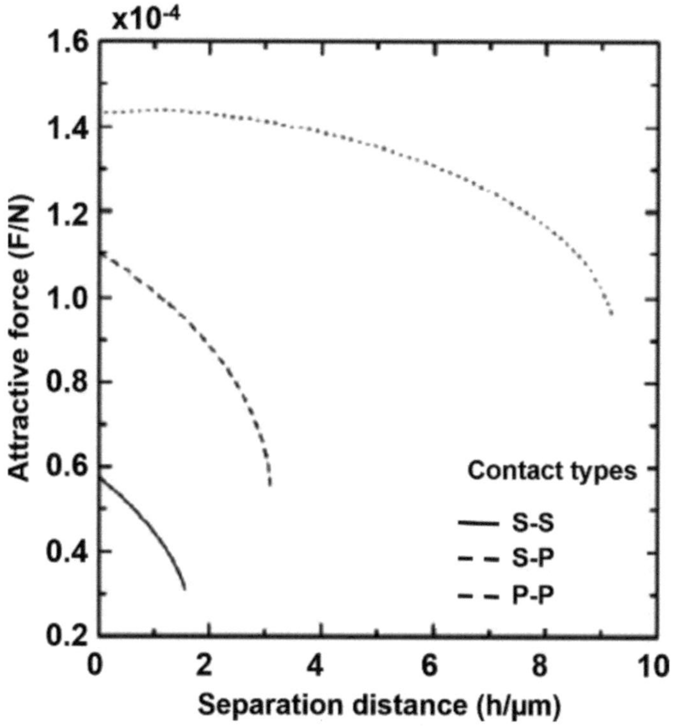

Figure 13.
Attractive force for different contact types. Zheng L, Malfliet a, Wollants P, Blanpain B, Guo M. effect of alumina morphology on the clustering of alumina inclusions in molten iron. ISIJ Int. 2016;56:926–935. DOI: 10.2355/isijinternational.ISIJINT-2015-561. Reproduced with permission [42].

In this equation, D_V = $1.3X10\, exp\left(\frac{-110000}{RT}\right)$ [43], $\sigma_{pp} = 1\frac{J}{m^2}$ is the interfacial tension between alumina particles K_1 = 10–100 [43]. **Figure 14** shows the mechanism of diffusion of vacancies in the alumina lattice to form the bond [44]. The bond strength reaches the order of MPa due to the interdiffusion between alumina particles [45].

3. Adherence of alumina particles on refractory surfaces

The adhesion of alumina inclusions to the refractory surface has its highest repercussion in the nozzle feeding with liquid steel the continuous casting slab mold. The property governing these phenomena is the contact angle between a particle and a solid phase with a smooth surface as is shown in **Figure 15**, which is a function of the surface tension and known as Young's law [47],

$$\sigma_{SL} = \sigma_{SV} - \sigma_{LV}\, cos\, \theta_Y \tag{22}$$

In actual refractory materials, there are not smooth surfaces and have certain levels of asperities and roughness. Hence, Eq. (22) suffers a modification through a roughness factor, r, to become in the Wenzel's Equation [47],

$$cos\, \theta_W = rcos\theta_Y \tag{23}$$

A further modification includes the consideration of the heterogeneous structural nature of industrial materials such as refractories for continuous casting and Eq. (23) becomes into the Cassie-Baxter Equation [47],

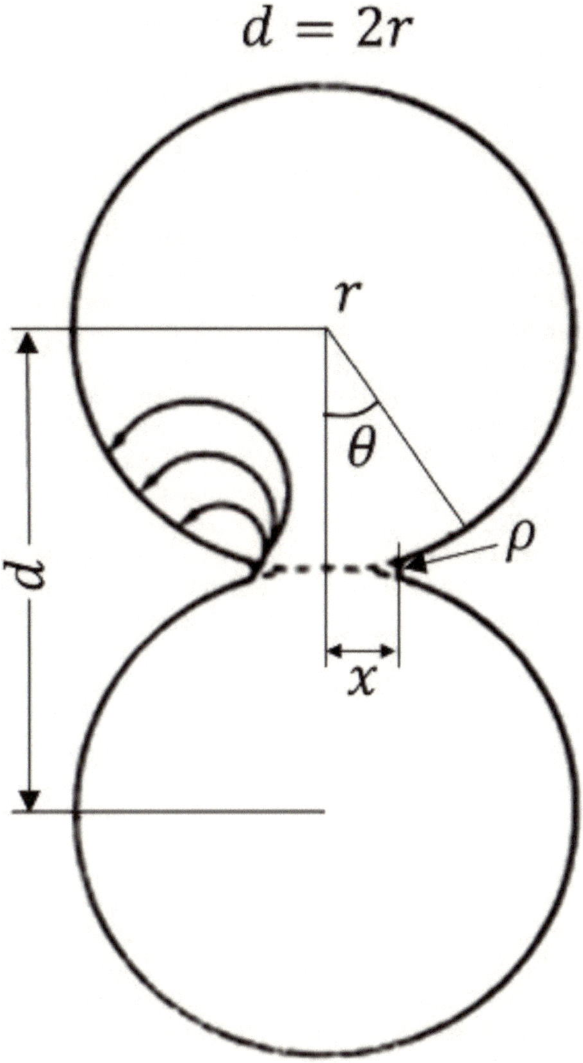

Figure 14.
Models for initial stages of sintering of spherical particles showing the vacancies diffusion paths [43].

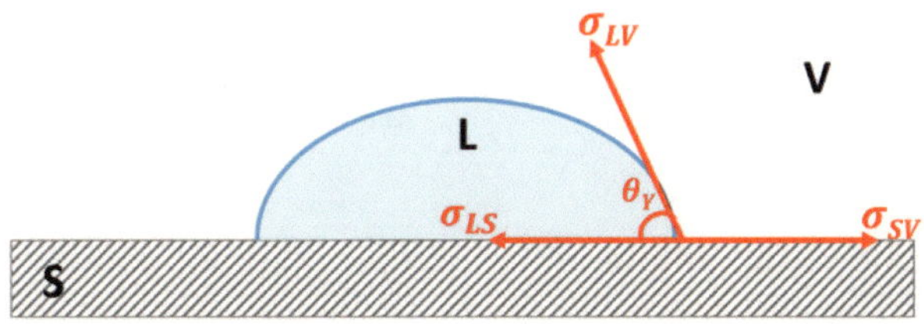

Figure 15.
Contact angle between surfaces in a liquid in equilibrium with a vapor phase [46].

$$\theta_{CB} = \frac{\sum_{i=1}^{n} f_n(\sigma_{i,SV} - \sigma_{i,SL})}{\sigma_{LV}} \tag{24}$$

Where the limits i and n, in Eq. (24), corresponding to the stable phases forming part of the refractory material and f_i is the surface fraction of phase i. The following conditions establish the wettability conditions: when $\sigma_{SV} > \sigma_{SL}$, $0^0 < \theta_{CB} < 90^0$, the liquid wets the solid. When $\sigma_{SL} > \sigma_{SV}$, $90^0 < \theta_{CB} < 180^0$, the solid and the liquid have a poor wettability. The work of adhesion between a particle and a substrate is derived from Young's Equation,

$$W_{ad} = \sigma_{SV} + \sigma_{LV} - \sigma_{SL} = \sigma_{LV}(1 + \cos\theta_{CB}) \tag{25}$$

A specific case of changes of surface tension of liquid steel is the case of ultra-low carbon steels stabilized with Ti (Ti-SULC steels) where this element decreases this property as seen in **Figure 16a**, and the work of adhesion as a function of the Ti content is shown in **Figure 16b** [46]. This effect of Ti on the surface tension of Ti-SULC steels enhances the wettability between the melt and the inclusions. The wettability of hydrophilic and hydrophobic systems (such as steelmaking and casting processes) increases and decreases, respectively, with surface roughness, leaving behind the ideal behavior, indicated by Young's Equation, as the best condition. **Figure 17a** shows the effects of the surface roughness on the contact angle or Wenzel angle. Increasing the roughness ratios make a hydrophilic system more hydrophilic and a hydrophobic system in a more hydrophobic one. It can be assumed that in the actual metal-refractory contact, due to poor wettability between the two phases, a gas can be trapped in between the asperities of the surface, such that the liquid sits on a surface having a distribution of solid asperities and gas pockets (two-component surface material) and their surface fractions are f_S and f_V respectively, where $f_S + f_V = 1$. Substituting in the CB Equation for the solid–liquid fraction $f_1 = f_S$ and $\cos\theta_{CB.1} = \cos\theta_{CB,2}$ and the gas pocket fraction $f_2 = f_V$ and $\cos\theta_{CB,2} = -1$ because the fraction is completely dry (no-wetting) and combining the roughness ratio factor r with the CB Equation, we get

$$\cos\theta_{app} = rf_S \cos\theta_{CB} + f_S - 1 \tag{26}$$

The examination of Eq. (26) indicates that if the fraction f_S approaches 0, by increasing the asperities of the surface, there will be a condition of perfect non-wettability. This trend is shown in **Figure 17b**. On the contrary, if f_S approaches 1 (complete surface smoothness), the contact angle is given by Eq. (26) with r = 1. The combined effects of roughness and a solid fraction are as follows: in a hydrophilic system, the simultaneous increases of solid fraction and surface roughness make a hydrophilic system more hydrophilic. A decrease of the solid fraction with a combined increase of surface roughness makes this hydrophobic system in a more

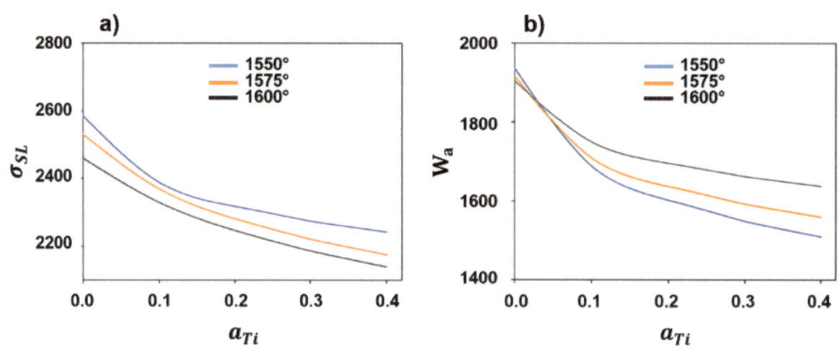

Figure 16.
Interfacial properties between liquid steel containing Ti and alumina particles, (a) interfacial tension, (b) adhesion work. González-Solórzano M.G., Morales R.D. Gutiérrez E, Guarneros J, Chattopadhyay K. analysis of fluid flow of liquid steel through clogged nozzles: Thermodynamics analysis and flow simulation. Steel. Res. Int. 2020;91. DOI: 10.1002/srin.202000049. Reproduced with permission [46].

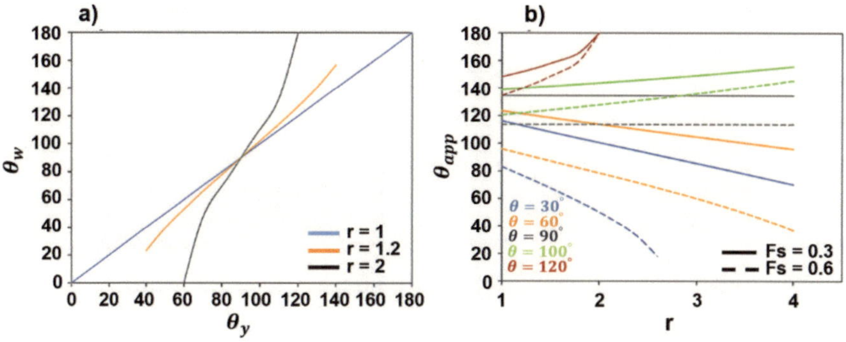

Figure 17.
Effect of roughness and voids on the surface refractory on interfacial properties, (a) comparison between Wenzel's and Young's equation, (b) effect of solid fraction for contact points according to the Cassie-Baxter equation. Of fluid flow of liquid steel through clogged nozzles: thermodynamics analysis and flow simulation. Steel. Res. Int. 2020;91. DOI: 10.1002/srin.202000049. Reproduced with permission [46].

hydrophobic one. Decreasing the adherence of inclusions on the refractory surface requires two simultaneous conditions that must be fulfilled: a small contact angle between the melt and the inclusion and a low contact angle between the refractory and the melt. Therefore, to manipulate the second angle, there may be two ways:

1. The first is to use CaO as a surface cover, which would be wetted it by liquifying the alumina inclusions and decreasing the contact angle between the melt and the calcium aluminate inclusion.

2. Use a conventional AG material with a special treatment leading to smooth surfaces according to the results presented in **Figure 17**.

The changes of wettability among the refractory, the melt, and the inclusions are summarized in **Figure 18(a–d)**:

a. The contact angle 1 (between the inclusion and the melt) is larger than the contact angle 2 (between the refractory and the melt). Hence, the liquid does not wet the nozzle. Therefore, the refractory rejects the metal, and there is the adhesion of the inclusion to the wall.

b. When the contact angle 2 is larger than angle 1, the nozzle is slightly wettable by the melt and allows that this one enters between the inclusion and the wall, making a small separation between them and reducing the strength of adhesion.

c. When angle 2 decreases further, the nozzle wall will increase its wettability by the melt. Hence, the separation between the inclusion and the wall becomes larger, making inclusion separate from the nozzle wall.

d. In this case, the liquid wets the inclusion; if the inclusion is liquid, it will go with the flow, but if it is solid, the inclusion approaches the wall and will remain adhered to it.

As the liquid steel flows through the nozzle, **Figure 19**, the alumina particles are transported along, and those close to the boundary layer might, eventually, get in this region and adhere to the refractory's surface by mechanisms of fluctuating velocities [48, 49]. Therefore, in current casting systems, the adherence of an

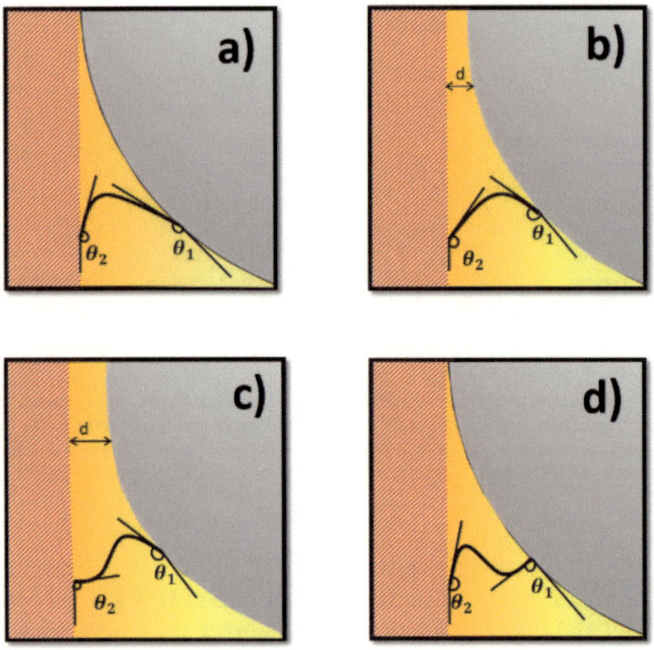

Figure 18.
Effects of melt-refractory and melt-inclusion contact angles on particle-refractory surface adhesion.

alumina particle to the surface of a nozzle refractory is favored by the low wettability of the inclusion and the refractory by the melt.

3.1 Adherence force

When a particle approaches the solid surface or to another particle, both unwetted by the liquid, the formation of a cavity between them becomes thermodynamically favorable, see **Figure 20**. This phenomenon is because the replacement of a particle-liquid interface by a particle-vapor interface leads to a negative change of the Gibbs free energy according to Young's Equation. Once the cavity is formed, an attractive force of adhesions develops, F_A, as expressed in the following Equation [50],

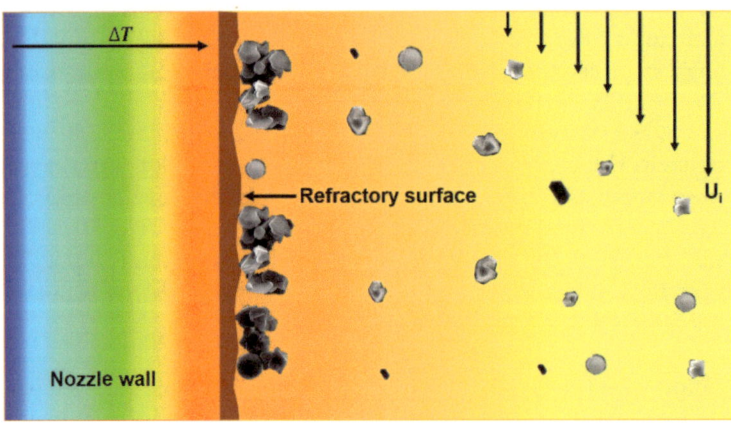

Figure 19.
Adhesion of alumina clusters and aggregates contained in the liquid flow on the nozzle surface.

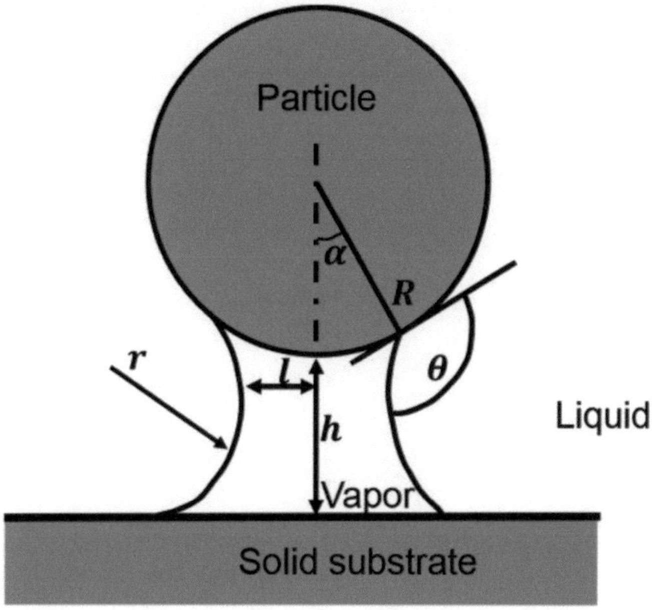

Figure 20.
Schematization of a cavity between a sphere and a plate in a non-wetting system.

$$F_A = 2\pi\sigma_{LV}l + \pi l^2 \Delta P \quad (27)$$

The cavity might be filled with: (1) gaseous components initially dissolved in the melt (2) gaseous components coming from the refractory (3) melt vapor, or (4) liquid phases forming due to a local rise in the oxygen concentration. In any case, there will be a pressure drop between the liquid outside of the cavity and the phase inside of it. The pressure difference obeys Laplace's Eq. (45),

$$\Delta P = \sigma_{LV}\left(\frac{1}{r} - \frac{1}{l}\right) \quad (28)$$

Where l and r are the principal radii of the cavity. The two parameters determine the profile of the liquid–vapor interface, which can be either optimized using the Laplace-Young Equation at constant pressure drop or approximately described with a piece of a circle with radius r. When the cavity is in equilibrium with the liquid phase, its shape is determined by thermodynamics. The estimated adhesion force between an alumina particle and the wall of an AG (alumina-graphite nozzle) is about $25X10^{-6}$ N [51], large enough to keep fixed the particle on the refractory surface.

4. Boundary conditions of wall adhesion

To design nozzle materials through the principles of physical-chemistry of interfaces and techniques of computer fluid dynamics, reliable boundary conditions for inclusion adherence to the refractory wall are necessary. In other words, to develop new materials is necessary to deal with a theoretical analysis including effects of surface roughness, the effect of impurities in alumina graphite materials (mainly Na_2O and K_2O, oxides cover) together with control of the boundary layer by the internal design of the nozzle [46] before making investments on experimental research. Specifically, this boundary condition is applicable only during the development of the clog's first layer and not during its growth. That is to say, when the first layer of the clog stabilizes, its future growth is guaranteed by contact with other alumina particles, which will go through a sinterization process by diffusion of vacancies, as explained above, forming high strength bonds. This boundary condition is derived from a balance of forces, according to **Figure 21.** These forces obey the following expressions [52]:

Drag force

$$F_d = 6\pi\mu\gamma\frac{d_p^2}{4}f \tag{29}$$

Adhesion force

$$F_a = \frac{3}{4}\pi\sigma d_p \tag{30}$$

Buoyancy force

$$F_b = \left(\rho_L - \rho_p\right)g\frac{1}{6}\pi d_p^3 \tag{31}$$

Lift force

$$F_L = 9.22\frac{\gamma\mu d_p^2}{4}\frac{\gamma d_p^2}{4v} \tag{32}$$

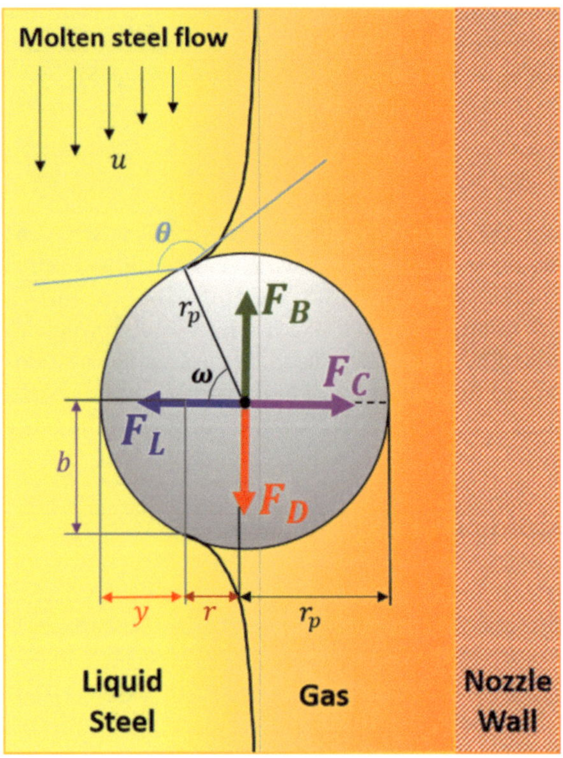

Figure 21.
Forces acting on an inclusion. (FC = capillary force, FB = buoyancy force, FL = lift force, FD = drag force) [51].

Different force ratios working on the particles are useful as possible boundary conditions or, simply, to compare the magnitudes of these forces as follows:

Ratio for the vertical lift-off

$$R_v = \frac{F_L}{F_a} \tag{33}$$

Ratio for the sliding:

$$R_s = \frac{F_d + F_b}{k_s(F_a - F_L)} \tag{34}$$

Ratio for the tangential lift-off:

$$R_t = \frac{(1.4r_p)F_d + aF_L + r_pF_b}{aF_a} \tag{35}$$

The constants and variables included in these expressions are reported in **Table 3** [53]. In these calculations, instead of using Eq. (30) to estimate the adhesion force, the force suggested of 25X10^{-6} N was employed [51]. This force was calculated by optimizing Eq. (27) and Eq. (28) and is considered a more realistic magnitude. The force ratios calculated for a ten μm particle in a liquid steel flow are in the last column of **Table 3**. Some conclusions derived from these results are a) The lift force is negligible compared with the adhesion force. b) The ratio for sliding is small compared with the effects of the adhesion force. c) Summing the momentums of drag, lift, and buoyancy forces and compared with the adhesion force's momentum yields a magnitude larger than one, meaning that the particle may be dislodged from the wall.

Closure

The clogging phenomena have their roots in the deoxidation step of the liquid steel. Operational factors like the addition time of aluminum, oxygen supersaturation, temperature, and melt stirring fixe the initial conditions of sizes distributions

Parameters for the calculation of ratios [53]		
Shear rate, γ	1 x 10^5 s^{-1}	**Ratio of forces**
Poisson's ratio, v_1	0.27	$R_v = 5.12x10^{-8}$
Poisson's ratio, v_2	0.28	
Poisson's ratio, E_1	8.01x10^{10}	
Poisson's ratio, E_2	2.15x10^{11}	
Surface energy, σ	0.001 J.m^{-2}	R_s =0.064
Coefficient of friction, k_s	0.3	
f	1.7009	
Viscosity, μ_l	0.006 Pa.s	
Liquid density, ρ_l	7100 kg/m^3	$R_t = 7.60$
Particle density, ρ_p	3600 kg/m^3	
Particle diameter, d_p	10 x 10^{-6} m	
Gravity, g	9.8 m/ s^2	

Table 3.
Parameters and constants to estimate the forces on a particle with a 10 μm diameter.

and the alumina particle's population responsible of the nozzle clogging. The high level of supersaturation required by aluminum governs the initial size distribution. Once the particle is nucleated at the initial stages, the initial growth is through diffusion and Ostwald ripening mechanisms. Although not proved yet here, the published literature reports that the bond strength among alumina particles and their morphology are important on the clogging mechanism. Once inside the nozzle, the particle, taking contact with refractory, may remain adhered to it if the adhesion force is larger than the momentums originated from the lift, buoyancy and drag forces with the particle size.

The refractory's roughness is of no help to control the clogging as those materials that are hydrophobic or hydrophilic will enhance these properties with rough surfaces. Ideally, a smooth surface approaching the Young's Law would be the ideal material to decrease clogging. Raw materials purity is of interest as some oxides are easily reduced by the carbon of the nozzle or aluminum in the melt, all working to enhance the clogging problem.

Under the present situation, this work contributes to the understanding of the surface phenomena in the areas of inclusion nucleation and growth of inclusions and the steel refining and the interaction with the refractory. It gives options for the boundary conditions applied in computational fluid dynamics simulations, all focused on designing new nozzle materials.

Nomenclature

C_O	Oxygen concentration in the melt
C_P	Oxygen concentration in the oxide
d_p	Particle diameter
D_O	Diffusion coefficient of oxygen in the melt
D_V	Coefficient diffusion of vacancies
f	Correction factor
F	Force
F_s, F_v	Surface fraction of solid and vapor phase, respectively
g	Gravity constant
h_i	Henryan activity of i
I	Nucleation rate
k_B	Boltzmann's constant
l	Principal radii
N_o	Constant
P	Pressure
r	Inclusion radii
r_0	Initial radii of the inclusion
r_c	Critical nuclei radii
R	Gas constant and particle radii
R_1	Particle 1 radii
R_2	Particle 2 radii
S_O	Supersaturation
t	Time
T	Temperature
V_O	Oxide molar volume
w	Collision probability of two particles in turbulent flow regime
W_{ad}	Work of adhesion
W_s	Probability for collisions for two particles in Stokes's regime
x	Neck radii

Greek letters

α	Accommodation factor
γ	Shear rate
ε	Dissipation rate of kinetic energy
θ	Contact angle
μ	Dynamic liquid viscosity
ν	Kinematic liquid viscosity
ρ	Density
σ	Surface tension

Sub indexes

a	Adhesion
app	Apparent
b	Buoyancy
c	critical
CB	Cassie-Baxter
d	Drag
het	Heterogeneous
L	Lift
LV	Liquid–Vapor
M	Metal
O	Oxide
PL	Particle-Liquid
SL	Solid–Liquid
SV	Solid-Vapor
v	Molar volume
w	Wenzel
y	Young

Author details

María-Guadalupe González Solórzano[1], Rodolfo Morales-Dávila[1*],
Jafeth Rodríguez Ávila[2], Carlos Rodrigo Muñiz-Valdés[2] and Alfonso Nájera Bastida[3]

1 Instituto Politécnico Nacional-ESIQIE, Department of Metallurgy, Mexico

2 Facultad de Ingeniería, Universidad Autónoma de Coahuila, Blvd. Fundadores Km 13, Ciudad Universitaria, 25350, Arteaga Coahuila, México

3 Instituto Politécnico Nacional-UPIIZ, Metallurgical Engineering, Blvd. del Bote 202, Cerro del Gato, 98160, Zacatecas, México

*Address all correspondence to: rmorales@ipn.mx

References

[1] Long M., Zuo X., Zhang L., Chen D., Kinetic modeling on nozzle clogging during steel billet continuous casting. ISIJ Int. 2010;50:712–720. DOI: 10.2355/isijinternational.50.712.

[2] Ni P., Jonsson L.T.I., Ersson M., Jönson PG., The use of an enhanced Eulerian deposition model to investigate nozzle clogging during continuous casting of steel. Metall. Mater. Trans. 2014;45:2414–2424. DOI: 10.1007/s11663-014-0145-5.

[3] Gutiérrez E., Garcia-Hernández S., Barreto J.deJ., Mathematical modeling of inclusions deposition at the upper tundish nozzle and the submerged entry nozzle. Steel Res. Int. 2016;87:1406–1416. DOI: 10.1002/srin.201500422.

[4] Gutiérrez E., Garcia-Hernández S., Barreto J.deJ., Mathematical analysis of the dynamic effects on the deposition of alumina inclusions inside the upper tundish nozzle. ISIJ Int. 2016;56: 1394–1403. DOI: 10.2355/isijinternational.ISIJINT-2016-076.

[5] Barati H., Wu M., Holzmann T., Kharicha A., Ludwig A., Simulation of non-metallic inclusion deposition and clogging of nozzle. Miner. Met. Mater. Ser. 2018:149–158. DOI: 10.1007/978-3-319-72059-3_15.

[6] Snow R.B., Shea J.A., Mechanism of erosion of nozzles in open-hearth ladles. J. Am. Ceram Soc. 1949;32:187–194. DOI: 10.1111/j.1151-2916.1949.tb19766.x.

[7] Braun T.B., Elliott J.F., Flemings M. C., The clustering of alumina inclusions. Metall. Mater. Trans. 1979;10:171–184. DOI: 10.1007/BF02652461.

[8] Miki Y., Kitaoka H., Sakuraya T., Fujii T., Mechanism for separating inclusions from molten steel stirred with a rotating electro-magnetic field. ISIJ Int. 1992;32:142–149. DOI: 10.2355/isijinternational.32.142.

[9] Zhang L., Thomas B.G., State of the art in the control of inclusions during steel ingot casting. Metall. Mater. Trans. B 2006;37:733–761. DOI: 10.1007/s11663-006-0057-0.

[10] Sasai K., Mizukami Y., Mechanism of alumina adhesion to continuous caster nozzle with reoxidation of molten steel. ISIJ Int. 2001;41:1331–1339. DOI: 10.2355/isijinternational.41.1331.

[11] Tehovnik F., Burja J., Arh B., Knap M., Submerged entry nozzle clogging during continuous casting of Al-killed steel. 2015. Available from: https://hrcak.srce.hr/128965 (Accessed: 2020-09-18).

[12] Fukuda Y., Ueshima Y., Mizoguchi S., Mechanism of alumina deposition on alumina graphite immersion nozzle in continuous caster. ISIJ Int. 1992;32:164–168. DOI: 10.2355/isijinternational.32.164.

[13] Sasai K., Mizukami Y., Yamamura H., Reaction mechanism between alumina graphite immersion nozzle and low carbon steel. J. Iron Steel Inst. Jpn. 1993;79:1067–1074. DOI: 10.2355/tetsutohagane1955.79.9_1067.

[14] Sasai K., Mizukami Y., Reaction mechanism between alumina graphite immersion nozzle and low carbon steel. ISIJ Int. 1994;34:802–809. DOI: 10.2355/isijinternational.34.802.

[15] Singh S.N., Mechanism of alumina buildup in tundish nozzles during continuous casting of aluminum-killed steels. Metall. Mater. Trans. 1974;5: 2165–2178. DOI: 10.1007/BF02643930.

[16] Kojola N., Ekerot S., Andersson M., Jönsson P.G., Pilot plant study of nozzle clogging mechanisms during casting of

REM treated stainless steels. Ironmak. Steelmak. 2011;38:1–11. DOI: 10.1179/030192310X12690127076398.

[17] Yokota S., Souma S.T., Guchi M., Asak Y., Hara S., Prevention of air suction from the contact-part between sliding gate and immersion nozzle. ISIJ Int. 1998;38:1346–1352. DOI: 10.2355/isijinternational.38.1346.

[18] Rackers K., Thomas B.G., Clogging in continuous casting nozzles. In: 78th Steelmaking Conf. Iron Steel Soc. April 2–5 1995; Nashville USA. p. 723–734.

[19] Savolainen J., Rousu A., Fabritius T., Mattila O, Sulasalmi P., Modelling of pressure distribution inside the SEN in a stopper-rod controlled system. Steel Res. Int. 2010;81:980–986. DOI: 10.1002/srin.201000012.

[20] Vermeulen Y., Coletti B., Blanpain B., Wollants P., Vleugels J., Material evaluation to prevent nozzle clogging during continuous casting of Al killed steels. ISIJ Int. 2002;42:1234–1240. DOI: 10.2355/isijinternational.42.1234.

[21] Barati H., Wu M., Kharicha A., Ludwig A., A transient model for nozzle clogging. Powder Techn. 2018;319:181–198. DOI: 10.1016/j.powtec.2018.01.053.

[22] Barati H., Wu M., Kharicha A., Ludwig A., Calculation accuracy and efficiency of a transient model for submerged entry nozzle clogging. Metall. Mater. Trans. B 2019;50:1428–1443. DOI: 10.1007/s11663-019-01551-x.

[23] Lee J.H., Kang Y.B., Growth of initial clog deposits during continuous casting of Ti-ULC steel formation and reduction of the initial deposits at nozzle/steel interface. ISIJ Int. 2020;60: 426–435. DOI: 10.2355/isijinternational. ISIJINT-2019-384.

[24] Lee J.H., Kang M.H., Kim S.K., Kang Y.B., Oxidation of Ti added ULC steel by CO gas simulating interfacial reaction between the steel and SEN during continuous casting. ISIJ Int. 2018;58:1257–1266. DOI: 10.2355/isijinternational.ISIJINT-2018-164.

[25] Lee J.H., Kang M.H., Kim S.K,. Kim J., Kim M.S., Kang Y.B., Influence of Al/Ti ratio in Ti-ULC steel and refractory components of submerged entry nozzle on formation of clogging deposits. ISIJ Int. 2019;59:749–758. DOI: 10.2355/isijinternational.ISIJINT-2018-672.

[26] Suito H., Ohta H., Characteristics of particle size distribution in early stage of deoxidation, ISIJ Int. 2006;46:33–41. DOI: 10.2355/isijinternational.46.33.

[27] Kubaschewski O., Alcock C.B., Metallurgical Thermochemistry, Pergamon Press. New York. 1979, 1.

[28] Tromp R.M., Hannon J.B., Thermodynamics of nucleation and growth, Surf. Rev. and Letters, 2002;9; 1565–1593.

[29] Li G., Suito H., Electrochemical measurement of critical supersaturation in Fe-O-M (M=Al, Si, and Zr) and Fe-O-Al-M (M=C, Mn, Cr, Si, and Ti) melts by solid electrolyte galvanic cell. ISIJ Int. 1997;37:762–769. DOI3: 10.2355/isijinternational.37.762.

[30] Van Ende Marie-Aline., Formation and morphology of non-metallic inclusions in aluminium killed steels [thesis]. Saint-Louis: Universite Catholique de Louvain, 2010.

[31] Wakoh M., Sano N., Behavior of alumina inclusions just after reoxidation, ISIJ Int. 2003;47;627–632.

[32] Porter D.A., Easterling KE, Sherif MY. Phase Transformations in Metals and Alloys, Chapman and Hall, 3er ed. London-Tokyo; 2009. 315 p. DOI: 10.1201/9781439883570.

[33] De Cooman B.C., Speer J.G., Fundamentals of Steel Product Physical

Metallurgy. 1st ed. Assn of Iron & Steel Engineers. 2011. 40 p.

[34] Kojima Y., Inouye M., Yamada Y., Solubility and diffusion of nitrogen in liquid iron-nickel and iron-cobalt alloys at 1600° C., Transactions ISIJ 1975;15: 599–605.

[35] Steinmetz E., Lindenberg H.U., Morphology of inclusions at aluminium deoxidation. Archiv Eisenhuttenwesen 1976;47:199–204.

[36] Adaba Menyelum-Obinn. Oxide Inclusion Evolution and Factors that Influence their Size and Morphology [thesis]. Missouri University of Science and Thechnology, 2019.

[37] Lindborg U., Torssell K., A collision model for the growth and separation of deoxidation products. Trans. Met. Soc. AIME, 1968; 242:94–102.

[38] Saffman P.G., Turner J.S., On the collision of drops in turbulent clouds. J. F!uid Mech. 1956;1:16–30.

[39] Miki Y., Kitaoka H., Sakuraya T., Fujii T., Mechanism for separating inclusions from molten steel stirred with a rotating electro-magnetic field. ISIJ Int. 1992;32:142–149. DOI: 10.2355/isijinternational.32.142.

[40] Dekkers R., Jokanovic N., Rombout A., Blanpain B., Wollants P., Steel cleanliness during secondary metallurgy of high-grade quality electric steels. Steel. Res. Int. 2005;76:475–480. DOI: 10.1002/srin.200506042.

[41] Delgado M., SIMEC Operations, private communication, Steelmaking Manager, SIMEC Group, Special Steels Plant. Tlaxcala, Mexico, 2019.

[42] Zheng L., Malfliet A., Wollants P., Blanpain B., Guo M., Effect of alumina morphology on the clustering of alumina inclusions in molten iron. ISIJ Int. 2016;56:926–935. DOI: 10.2355/isijinternational.ISIJINT-2015-561.

[43] Ooi H., Sekine M., Kasai G., On the Mechanisms of alumina cluster formation in molten iron. Tetsu-to-Hanage 1973;59:1078–1088. DOI: 10.2355/tetsutohagane1955.59.8_1078.

[44] Kingery W.D., Bowen H.K., Uhlmann D.R., Introduction to Ceramics. 2nd ed. Wiley; 1976. 474–475 p.

[45] Kuczynski G.C., Selft-diffusion in sintering of metallic particles. Trans. Met. Soc. AIME 1949;185:169–178. DOI: 10.1007/978-94-009-0741-6_33.

[46] González-Solórzano M.G., Morales R.D., Gutiérrez E., Guarneros J., Chattopadhyay K., Analysis of fluid flow of liquid steel through clogged nozzles: thermodynamics analysis and flow simulation. Steel. Res. Int. 2020;91. DOI: 10.1002/srin.202000049.

[47] Eustathopoulos N., Nicholas M.G., Devrit B., Wettability at High Temperatures. 1st ed. Pergamon Press: Amsterdam; 1999. 1 p.

[48] Suito H., Ohta H., Characteristics of particle size distribution in early stage of deoxidation. ISIJ Int. 2006;46:33–41. DOI: 10.2355/isijinternational.46.33.

[49] Rashidi M., Hestroni G., Barnejee S., Particle-turbulence interaction in a boundary layer. Int. Journal of Multiph. Flow 1990;16:935–949. DOI: 10.1016/0301-9322(90)90099-5.

[50] Simons S.J.R., Seville J.P.K., Adams M.J., Analysis of the rupture energy of pendular liquid bridges. Chem. Eng. Sci. 1994;49:2331–2339. DOI: 10.1016/0009-2509(94)E0050-Z.

[51] Diéguez-Salgado U., Investigation of particle attraction by steel/refractory and steel/gas interfaces and the associated relevance for clogging in casting processes [thesis]. Leoben; University, Austria, 2018.

[52] Liming Shi L., Bayless D., Comparison of boundary conditions for

predicting the collection efficiency of cyclones Pow. Tech. 2007;173:29–37. DOI: 10.1016/j.powtec.2006.11.022.

[53] Johnson K.L., Kendall K., Roberts A. D., Suface energy and the contact of elastic solids. In: Proceedings of the Royal Society A. 08 September 1971. p. 301–313. DOI: 10.1098/rspa.1971.0141.

Chapter 7

Melt Treatment-Porosity Formation Relationship in Al-Si Cast Alloys

Dominique Gagnon, Agnes M. Samuel, Fawzy H. Samuel, Mohamed H. Abdelaziz and Herbert W. Doty

Abstract

The present study focuses on the porosity formation in three Al-Si cast alloys widely used in automotive industries viz. A319.0, A356.0, and A413.0 alloys under various conditions: stirring, degassing. Sr level, amount of grain refining, combined modification and grain refining, as well as hydrogen level. The solidification rate was the same for all alloys in terms of the mold used and its temperature. The microstructural investigations were carried out quantitatively using an optical microscope-image analyzer system scanning systematically over a polished sample area of 25 mm × 25 mm and qualitatively using an electron probe microanalzer equipped with EDS and WDS systems. The results were coupled with hardness measurements.

Keywords: aluminum alloys, modification, grain refining, degassing, oxides, porosity

1. Introduction

The presence of porosity in aluminum castings is inevitable to a certain extent. Porosity in aluminum-silicon castings occurs when dissolved hydrogen liquid metal is rejected out during solidification through the interdendritic regions to compensate for the shrinkage associated with the solidification caused by the change in the casting volume. Hydrogen is the only gas that dissolves to a large extent in molten aluminum, leading to the formation of porosity [1, 2]. Majority of the studies of porosity formation in aluminum alloys have been carried out on aluminum silicon cast alloys [3–6]. According to Bian et al. [7], porosity can be classified into two categories: (i) microporosity (1–10 mm). This type of porosity is mainly shrinkage due to lack of molten metal feeding which is normally observed in alloys with long freezing range (e.g. 319 and 356 alloys), and (ii) microporosity (<500 μm) distributed uniformly throughout the matrix and caused by the rejection of the dissolved hydrogen gas.

Beside the hydrogen, modification with Sr or grain refining using Al-Ti-B master alloys would influence the formation of porosity. The porosity formed in Sr-modified castings has been related to a number of parameters i.e. hydrogen level of the melt, feedability (long mushy zone), and changes in the mechanism of eutectic nucleation [8]. Porosity characteristics in Sr-treated alloys depend on the amount of Sr oxides formed during solidification. Also, the presence of aluminum oxide films would result in the formation of large pores that are linked together. The end result can be explained in terms of the difference in eutectic solidification in unmodified and Sr-modified alloys [9, 10]. Dinnis et al. [11] examined the amount, distribution, and morphology of porosity in sand-cast plates of Sr-free and Sr-containing Al-(0–9 wr%) Si alloys. The authors found that no apparent differences in the amount, distribution, and morphology of porosity were observed between Sr-free and Sr-containing alloys with no Si (i.e. pure Al). However, Sr modification significantly changed the amount, distribution, and morphology of porosity in alloys with high Si content.

Influence of oxides on porosity formation in Sr-treated alloys was investigated by Samuel et al. [12] in 319 and 356 alloys. Their findings show that porosity formation is frequently associated with strontium oxides (particles or films), as well as β-Al_5FeSi phase platelets. The Al_2SrO_3 oxides are formed during melting by the high oxygen affinity of strontium. These oxides are difficult to be removed via degassing. Aluminum oxide films trapped in the molten metal result in the precipitation of coarser and deeper pores than those formed due to the strontium oxides [13].

According to Campbell and Tiryakioğlu [14], the beneficial effect of Sr as a modifying agent could be compromised by porosity development caused by the formation of oxide bifilms. The authors suggested an explanation in terms of the oxide population in the melt. A new multi-zone model is proposed by SkatTiedje et al. [15]. This model was developed by considering the effect of cooling rate on solidification and distribution of porosity in cast Al–Si alloys. The effect of Sr-modification on pore formation in reduced pressure and atmospheric conditions was studied by Miresmaeili et al. [16–18]. A new type of metallic cup with a riser was designed for use in a reduced pressure test as a mold. It was designed to improve directional solidification so that no macro shrinkage occurred in the test samples.

Another parameter to be considered is the presence of Fe-based intermetallics, in particular β-Al_5FeSi. Taylor et al. [19] investigated the role of iron in the formation of porosity in Al-Si-Cu-based casting alloys. Their analysis pointed to inadequacies in the existing theories regarding the role of iron and suggests that a new theory is required to understand the observed behavior. Effect of alloy composition on the stability of the β-AlFeSi phase and the role of the latter in porosity formation in 319 and 356 alloys was studied by Khalifa et al. [20]. The β-phase particles are potential sites for porosity formation regardless of the alloy composition and the type or size of the β particles. SrO and Al_2Si_2 Sr particles were observed to have contact with the β-AlFeSi platelets in the microstructure. Thus, with the addition of Sr, finer, better distributed pores appear in the microstructure. The alloy tensile properties are greatly improved by: (a) Sr addition in the range 200–400 ppm, (b) increasing the Mn/Fe ratio to 0.7, (c) addition of 0.08%Be, (d) addition of 0.08% Be+0.02% Sr [21–23].

The present study was undertaken to emphasize the role of melt treatment parameters (stirring, degassing, Sr-modification, grain refining, hydrogen content, intermetallics) as well as alloy composition on porosity formation in Al-(6–11 wr%Si) cast alloys and their impact on the alloy hardness. The initial alloys were unmodified and contained 0.0075%Ti and less than 0.2%Fe.

2. Experimental procedure

Table 1 lists the chemical composition of the as received ingots. The ingots of each alloy were cut, cleaned and melted in an electrical heated furnace using 6 kg SiC crucibles. The melting temperature was about 750°C ± 5 °C. Degassing was

Alloy	Cu	Mg	Mn	Si	Fe	Sr	Ti	B
A319.0	3.112	0.0071	0.0005	5.59	0.170	0.0000	0.0105	0.0004
A356.0	0.060	0.314	0.0005	6.21	0.099	0.0001	0.0002	0.0002
A413.0	11.17	0.046	0.218	11.17	0.204	0.0000	0.0652	0.0003

Table 1.
Chemical composition of the as-received alloys (wt.%).

carried out using a perforated graphite impeller turning at 135 rpm using pure Ar gas (0.492 cubic meter/hr) for about 30–40 minutes. The same impeller was also used as stirrer without gas. Strontium was added in the form of Al-10%Sr master alloy (waffles) whereas grain refiner was introduced as Al-5%Ti-15B master alloy in the form of short rods. It should be mentioned here that the manufacturing of the grain refiner produces Al_2O_3 which will increase in the amount of porosity. For this reason, the grain refiner in this study was added in the form of short rods (cold rolled) to minimize the presence of such oxides.

Due to the high affinity of Sr to react with oxygen to form SrO oxides, Sr was added during the last 10 minutes of degassing. As for varying the hydrogen level, small pieces of raw potato were added at the end of degassing, (in this case, Sr was not used) followed by AlScan™ measurement. In addition, reduced pressure test (RPT) samplings were taken from the melts. **Figure 1** shows examples of RPT samples sectioned in half to examine the porosity observed under different melt conditions. In all cases, samples for chemical analysis were also obtained from each melt prior to pouring.

Following melt treatment, the molten metal was poured in L-shaped metallic molds (75 mm × 35 mm × 30 mm). The molds were coated with fine grained boron nitride and were preheated at 450°C **Figure 2**. **Table 2** summarizes the sample codes and their melt treatment whereas **Table 3** shows complete analysis of A356.0 alloy as an example.

Samples for metallographic examination were sectioned from the middle of the cast bars (25 mm × 25 mm) and polished following standard procedures [24]. Porosity characteristics were evaluated using a Clemex image analyzer in conjunction with an optical microscope. Measurements were carried out at 100× as shown in **Figure 3**. For each sample at least 100 fields were scanned to enhance the accuracy of measurements. Selected samples were examined using an electron probe microanalyzer (EPMA) equipped with EDS and WDS systems, operating at 20 kV.

In order to quantify the effect of melt treatment parameters, hardness measuremnts were carried out on the sample surface shown in **Figure 2** using Brinell hardness testing machine using a 10 mm diameter steel ball and 500 kgf force. For each condition, an average of 8 readings was reported. The samples of A319.0

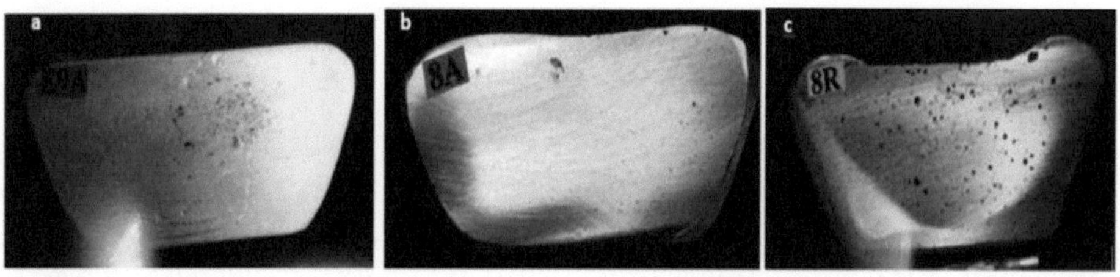

Figure 1.
RPT test taken from: (a) as received alloy, (b) after degassing, (c) 0.25 ml/100 g H_2.

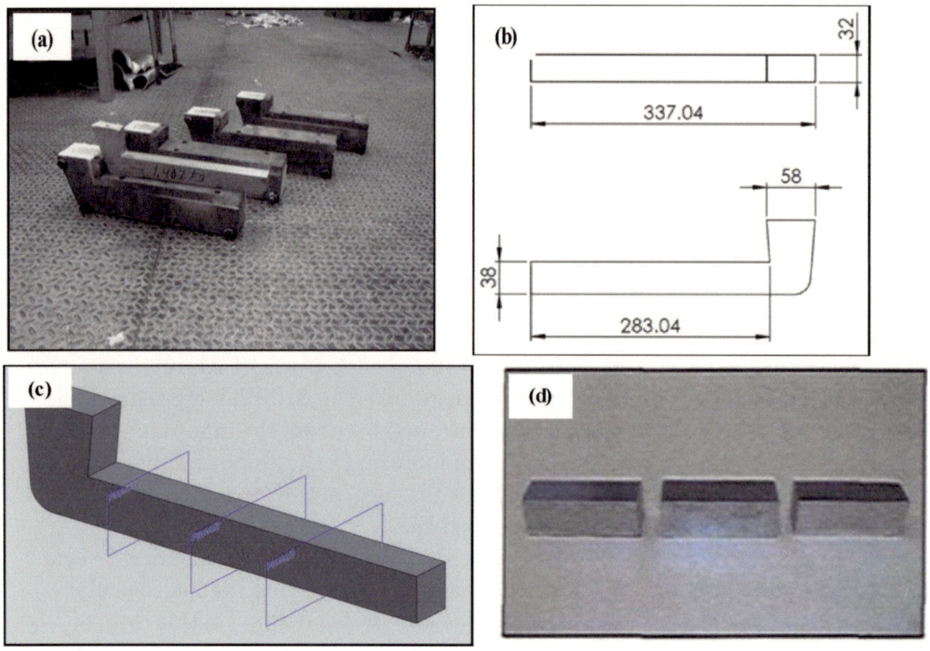

Figure 2.
L-Shape castings: (a) L-shape mold, (b) geometry of the L-shape casting, (c) cutting sequence of L-shaped casting to produce smaller rectangular bars, and (d) hardness test bars.

Metallurgical parameter	A319.0	A356.0	A413.0
Without degassing	A1	B1	C1
After degassing	A2	B2	C2
WithTiB$_2$ (0.11% Ti)	A3	B3	D3
With Sr (0.02%)	A4	B4	D4
With TiB$_2$ (0.11% Ti) and Sr (0.02%)	A5	B5	D5
H2 (0.15 ml/100 g)	A6	B6	D6
H2 (0.15 ml/100 g)	A7	B7	D7
With TiB$_2$ (0.11% Ti)	A8	B8	D8
With TiB$_2$ (0.22% Ti) and Sr (0.02%)	A9	B9	D9

Table 2.
Codes of the used alloy s and their melt treatment.

alloy were solution heat treated at 510°C for 8 h followed by water quenching. The samples thereafter were aged at various temperatures for 5 h followed by air cooling.

3. Results and discussion

3.1 Grain size

Selected samples from three levels of grain refining were used for macrostructure analysis. The polished samples were etched in a solution composed of (66% HNO_3, 33% HCl, 1% HF) and results are listed in **Table 4**. It is evident that with the increase in the amount of the added TiB_2 there is a significant decrease in the alloy

Code	Element (wt%)										
	Si	Fe	Cu	Mn	Mg	Cr	Ti	B*	Sr	H_2	Al
B1	6.24	0.0993	0.0574	0.0005	0.3137	0.0004	**0.0002**	<0.0002	**0.0001**	degassed	93.3
B3	6.21	0.0991	0.0597	<0.0005	0.3143	<0.0005	**0.1075**	<0.0002	**0.0001**	degassed	93.2
B3	6.29	0.0927	0.0067	<0.0005	0.3171	<0.0005	**0.1127**	<0.0002	**0.0001**	degassed	93.1
B5	6.16	0.0943	0.0032	<0.0005	0.3149	<0.0005	**0.1082**	<0.0002	**0.0159**	degassed	93.3
B5	6.09	0.0925	0.0066	<0.0005	0.3121	<0.0005	**0.1122**	<0.0002	**0.0193**	degassed	93.3
B6	6.17	0.1781	0.2087	0.0107	0.2854	0.0007	**0.2104**	0.0360	**0.003**	degassed	92.6
B7	5.96	0.1596	0.1992	0.0101	0.2716	<0.0005	**0.2307**	0.0360	**0.0172**	degassed	92.8
B8	6.10	0.0990	0.1124	<0.0005	0.3116	<0.0005	**0.1094**	<0.0002	**0.0002**	~0.15 ml/100 g	93.2
B9	6.18	0.1052	0.0887	0.0011	0.3102	0.0007	**0.1117**	<0.0002	**0.0001**	~0.25 ml/100 g	93.1

*B was not determined.

Table 3.
Actual chemical composition of A356.0 alloy melts prepared.

grain size, reaching about 80% when the concentration of grain refiner is about 0.2%Ti. **Figure 4** reveals the significant reduction in the grain size of the A413.0 alloy refined with 0.22%Ti. EPMA examination indicated that the initial Al-Ti-B master alloy was decomposed into TiB_2 (in the form of ultra-fine dispersed particles

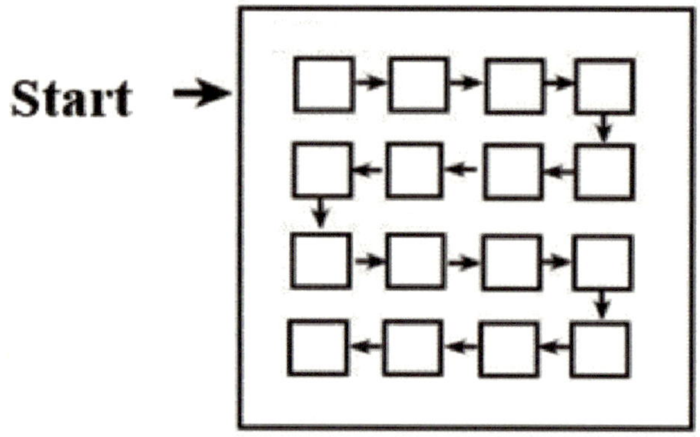

Figure 3.
Schematic distribution of the examined fields.

Code	Grains size (μm)	SD (μm)	%Refining
A1	678	29	base
A3	544	248	20
A6	214	54	68
B1	811	178	base
B3	692	293	15
B6	229	78	71
D1	2466	386	base
D3	2156	269	12
D6	475	97	80

Table 4.
Average grain size of the three alloys used.

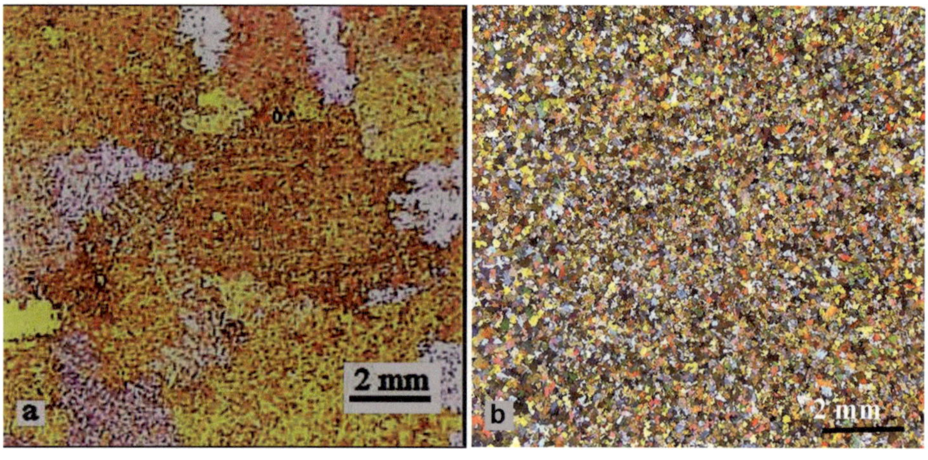

Figure 4.
Reduction in the grain size of the A413.0 alloy: (a) 0.007%Ti, (b) 0.22%Ti.

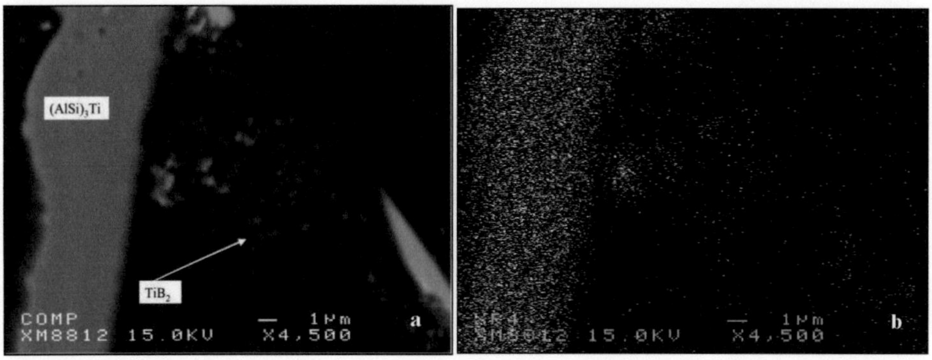

Figure 5.
Decomposition of the added Al-5%Ti-1%B master alloy: (a) backscattered electron image, (b) X-ray image of Ti distribution.

(50–100 nm) causing the observed refining) and platelets of $(AlSi)_3Ti$ within the α-aluminum as demonstrated in **Figure 5** [25, 26].

3.2 Porosity

It is deemed important to present the solidification curves (0.8°C/s) of the three used alloys in order to arrive at a clear understanding of the parameters that control the porosity characteristics. These curves are shown in **Figure 6**.

It is well established that porosity is formed due to dissolved hydrogen in the liquid metal, shrinkage during solidification process, and oxide films. Severity of porosity depends on the amount of these three parameters [1–3]. **Tables 5–7** present the measurements of the porosity characteristics as a function of the applied melt treatment. Degassing for a sufficient amount of time would lead to reduction in the dissolved hydrogen in the liquid metal as well as removal of most of inclusions and oxide films [14–18]. Thus all alloy melts were degassed prior to casting except for the hydrogen–containing alloys (coded 6 and 7) where degassing was done prior to introduction of H_2. It is inferred from these tables that hydrogen results in precipitation of large pores leading to a relatively high percentage of porosity with low pore density. Although Sr was added only 10 minutes before the end of degassing, however, the SrO formed during this period was enough to increase the amount of porosity but not as high as that caused by H_2.

The combined effect of oxides associated with the manufacturing of the grain refiner (mainly Al_2O_3) along with SrO contributed to the increase in the amount of porosity. It is noticed that the porosity in the A413.0 alloy is relatively less than in the other two alloys which may suggest that as-received alloy was previously degassed using chlorine, as inferred from its low Mg content (see **Table 1**). Also, it is should be taken into consideration that A413.0 alloy has a short freezing range compared to A319.0 and A356.0 alloys, see **Figure 5**. The average aspect ratio of the samples without degassing falls in the range of 2.0–2.3. After gassing, the average values of the aspect ratio lie in the range 1.2–1.32 indicating that the precipitated pores are more-or-less of spherical shape. However, in all cases, degassing with Ar could reduce the porosity by about 90%.

Figure 7 presents a comparison of average pore area and porosity percentage parameters for the three used alloys, highlighting the effect of oxides (SrO, Al_2O_3 films and/or bifilms), and the effect of hydrogen content. As can be seen there is a direct proportionality between the two parameters, associated with low pore density indicating porosity agglomeration which would lead to poor mechanical properties.

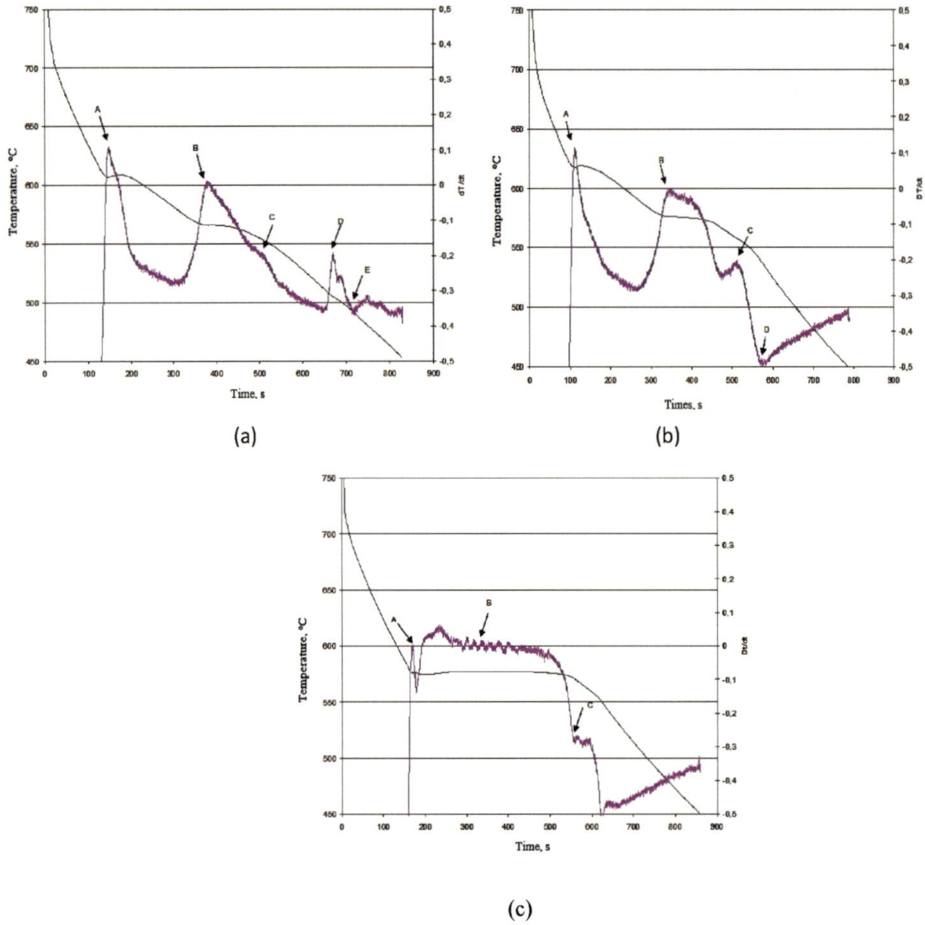

Figure 6.
Solidification curves and their first derivatives obtained for: (a) A319.0, and (b) A356.0, (c) A413.0 alloys.

Code	Average pore area (μm^2)		Average pore length (μm)		Aspect ratio		Density (pores/μm^2)	Ave area percentage (%)	
	Ave.	SD	Ave.	SD	Ave.	SD	—	Ave.	SD
A1	682	231	41	16	2.3	0.70	5.86	0.356	0.207
A2	99	123	12	08	1.63	0.69	2.80	0.029	0.018
A3	129	61	25	17	1.53	0.76	5.89	0.125	0.108
A4	844	119	30	17	1.60	0.56	2.78	0.780	0.041
A5	762	371	75	25	1.86	0.69	9.89	0.488	0.135
A6	975	325	40	12	1.32	0.62	6.64	1.845	0.317
A7	1505	368	106	23	1.29	0.53	5.61	2.414	0.039
A8	465	100	36	21	1.78	0.61	9.66	0.418	0.215
A9	1023	448	43	14	1.78	0.68	7.05	0.843	0.267

Table 5.
Characteristics of porosity in A319.0 alloy.

In general, the formation of pores is governed by the relation: $\Delta P = 2\sigma/r$, where σ is the surface tension, and ΔP is the critical pressure needed to create a nuclei of radius r [27]. Thus, introducing a high level of H_2 in the liquid

Code	Average pore area (μm^2)		Average pore length (μm)		Aspect ratio		Density (pores/μm^2)	Ave. area percentage (%)	
	Ave.	SD	Ave.	SD	Ave.	SD	—	Ave.	SD
B1	201	72	74	43	2.01	0.52	3.22	0.570	0.455
B2	59	81	14	11	1.83	0.43	2.08	0.026	0.049
B3	178	325	19	19	1.67	0.84	2.58	0.168	0.052
B4	1113	208	48	24	1.35	0.65	1.30	0.424	0.078
B5	1014	129	123	46	1.24	0.85	4.61	0.604	0.170
B6	1230	612	54	22	1.29	0.39	1.17	1.451	0.283
B7	1893	189	98	34	1.15	0.40	2.89	1.952	0.196
B8	484	274	26	11	1.75	0.68	2.75	0.388	0.060
B9	1488	452	29.76	24.77	1.70	0.69	5.78	0.556	0.126

Table 6.
Characteristics of porosity in A356.0 alloy.

Code	Average pore area (μm^2)		Average pore length (μm)		Aspect ratio		Density (pores/μm^2)	Ave area percentage (%)	
	Ave.	SD	Ave.	SD	Ave		Ave.	Ave	SD
D1	157	244.9	15	08	2.12	0.38	0.17	0. 20	0.004
D2	82	71.44	15	09	1.76	0.62	0.19	0.01	0.002
D3	454	462	30	14	1.56	0.65	0.83	0.134	0.038
D4	832	236	40	16	1.56	0.55	2.19	0.357	0.222
D5	935	296	61	21	1.87	0.60	1.14	0.592	0.148
D6	1490	341	17	21	1.31	0.53	1.72	0.643	0.157
D7	1840	802	46	21	1.29	0.41	2.08	0.857	0.186
D8	785	122	19	07	1.67	0.75	5.97	0.305	0.066
D9	965	181	35	05	1.70	0.66	6.58	0.552	0.182

Table 7.
Characteristics of porosity in A413.0 alloy.

aluminum would increase ΔP, decreasing r, and hence increasing the porosity percentage. **Figure 8(a)** shows an example of the pore shape observed in gassed A413.0 alloy (coded D6) where the pore is almost round. **Figure 8(b)** represents a pore caused by the shrinkage of the casting on going from liquid to solid state. Due to volume change associated with this process, the final casting shrinks by about 7%, which is accommodated within the final cast in the form of irregular pores passing through the interdendritic structure (**Figure 8(b)**). In certain cases, the gas pore could later on change into a shrinkage pore as seen in **Figure 8(c)**, depending on the level of dissolved gas.

Formation of oxide films during the course of melting due to insufficient degassing could lead to formation of massive pores as in the case of Sr-modified alloys where Sr was exposed to the outer atmosphere for a long period of time, as exemplified in **Figure 8(d)**. Li et al. [28] determined the nature of this oxide as $Al_{2.3}SrO_{3.3}$ using WDS analysis. Another type of oxide films associated with the pores is displayed in **Figure 8(e)** where the molten metal was partially stirred. According to Mohanty et al. [29] this type of oxide can act as nucleation sites for pore formation

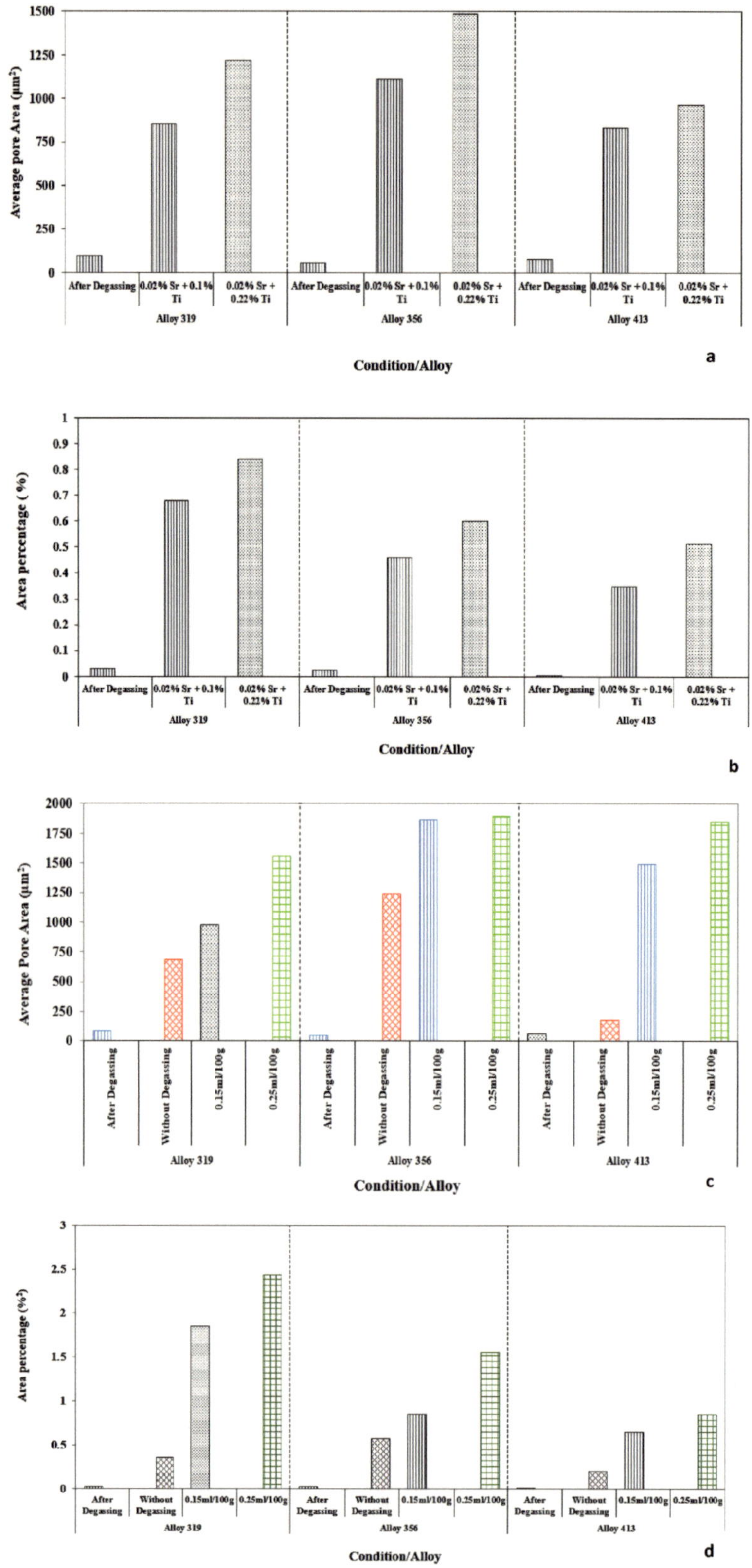

Figure 7.

Comparison of oxides and hydrogen level on the alloy pore size and porosity percentage: (a, b) effect of oxides as a function of the total amount of added Sr and grain refiner, (c, d) effect of hydrogen level.

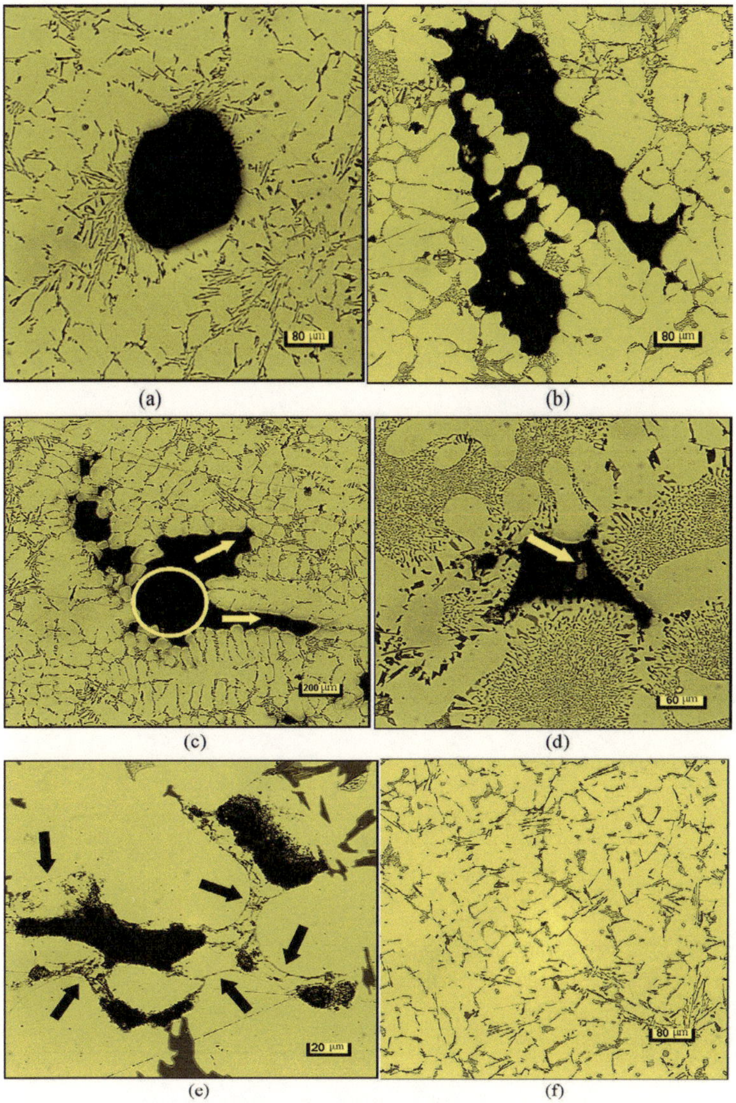

Figure 8.
Effect of melt treatment on the pore morphology in A413 alloy: (a) D6, (b) D9, (c) D7, (d) D3, (e) D1, (f) D2.

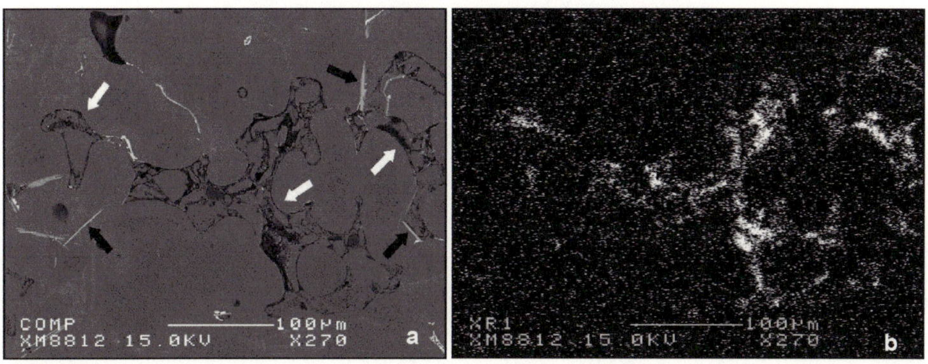

Figure 9.
(a) Backscattered electron image of D9 alloy, (b) X-ray map shownning distribution of O_2 in (a).

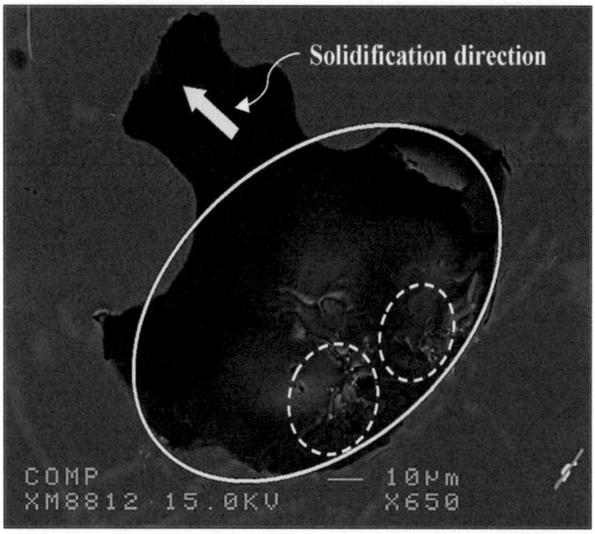

Figure 10.
Backscattered electron image revealing a massive pore composed of gas and shrinkage.

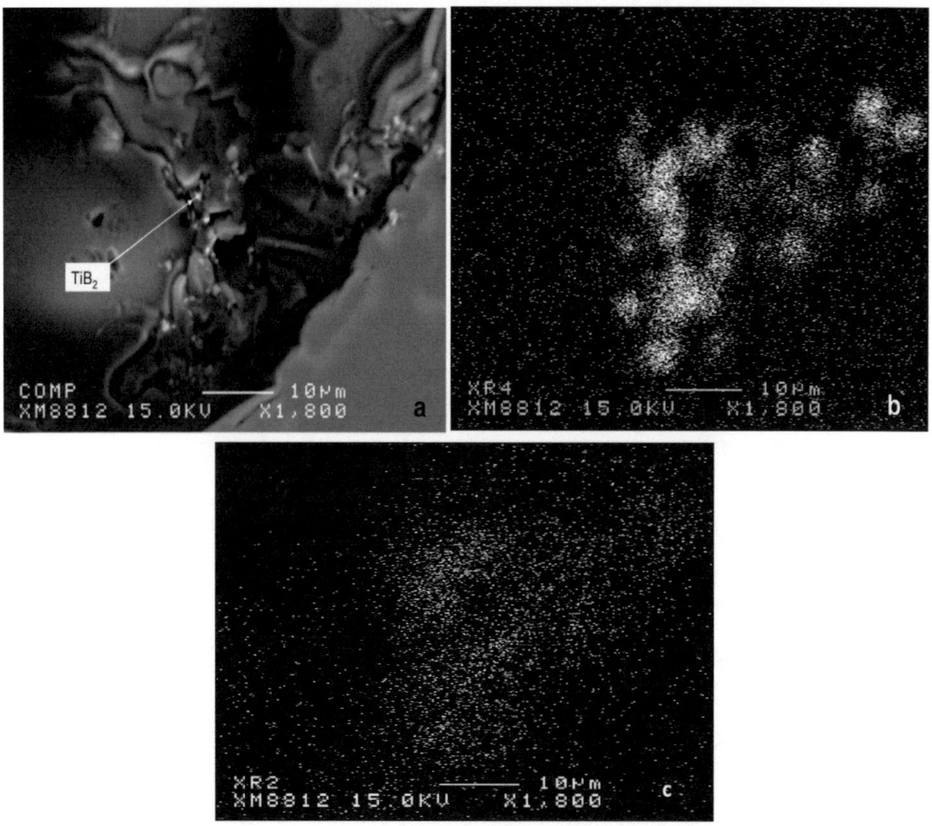

Figure 11.
(a) High magnification backscattered electron image of the white area in ***Figure 7*** *(broken line), (b, c) X-ray maps revealing the distribution of Ti and B, respectively, in (a).*

as shown by the black arrows in **Figure 7(e)**. **Figure 8(f)** exhibits the microstructure of a well degassed sample. **Figure 9(a)** is a backscattered electron image of D9 alloy (high Ti content) revealing the oxide films or bifilms [14] associated with the

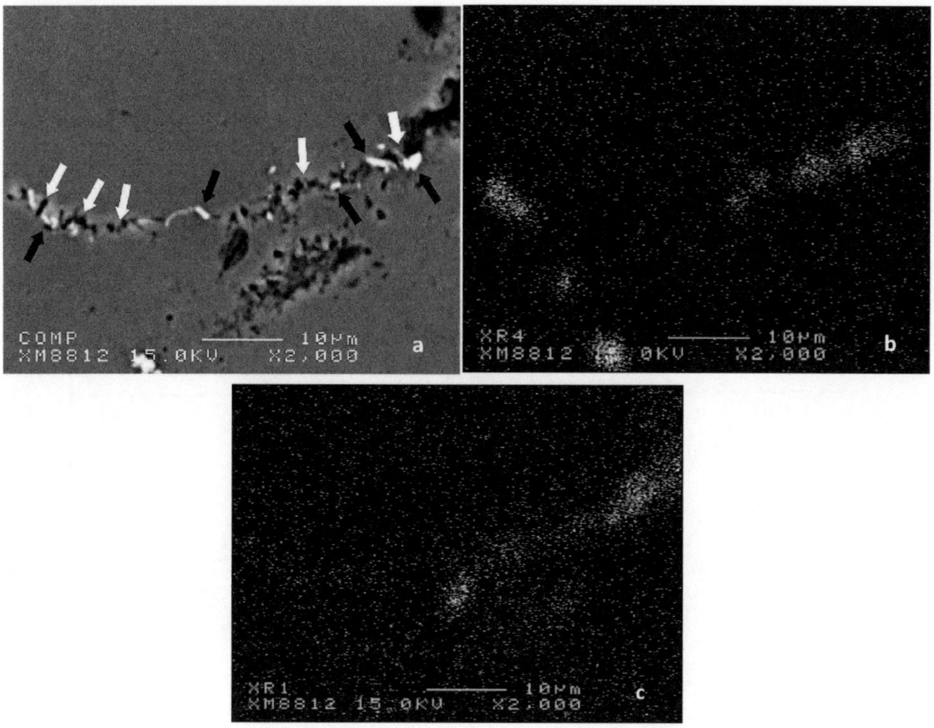

Figure 12.
(a) Backscattered electron image of TiB_2 particles mixed with SrO particles/films in D9 alloy, (b and c) X-ray maps of Ti and O_2, respectively in (a). The bright spots are fragments of SrO films.

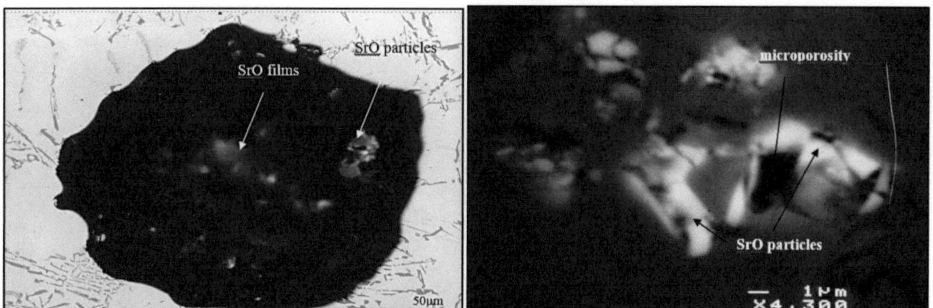

Figure 13.
(a) SrO films and particles observed within a pore in 319 alloy; (b) backscattered electron image showing SrO particles situated inside the pore [30, 31].

manufacturing of the Al-Ti-B master alloy (white arrows) and present in the sample due to insufficient degassing or mechanical stirring. The black arrows highlights the presence of $(Al.Si)_3Ti$ platelets in the vicinity of the oxide films.

Figure 10 is a backscattered electron image of **Figure 8(c)** displaying the change in the nature of the precipitated pore from round (due to H_2) into irregular (shrinkage-white arrow) pore during the coarse of solidification leading to the formation of a massive pore in D9 that contains high amounts of Ti and Sr. The presence of utlra fine particles was noticed within the gas pore (white broken lines) in **Figure 10**. A high magnification image of this part is displayed in **Figure 11(a)** revealing the presence of

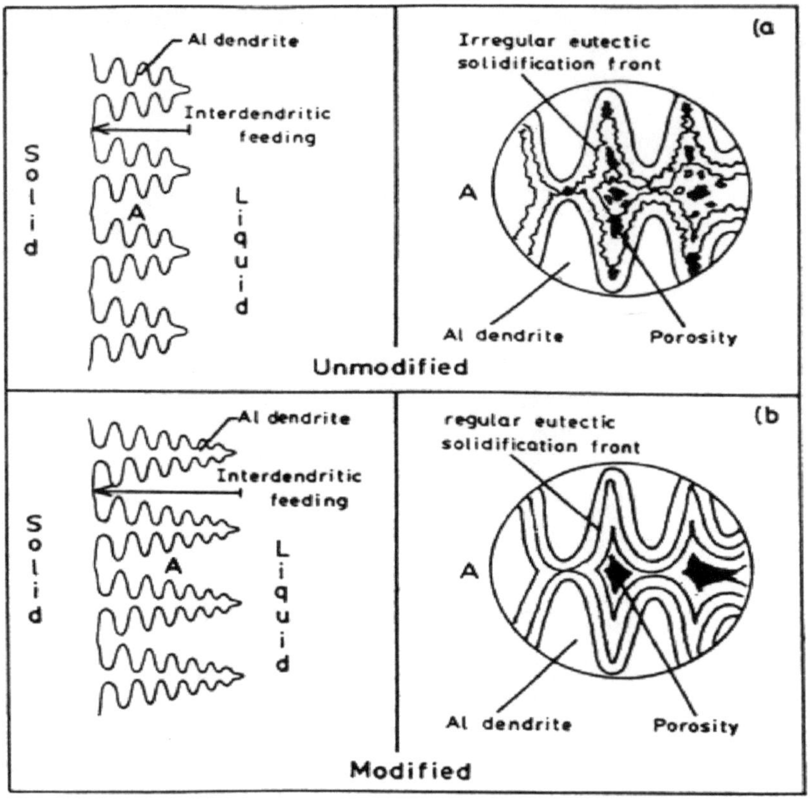

Figure 14.
The formation of porosity in: (a) unmodified alloys, (b) Sr-modified alloys [32].

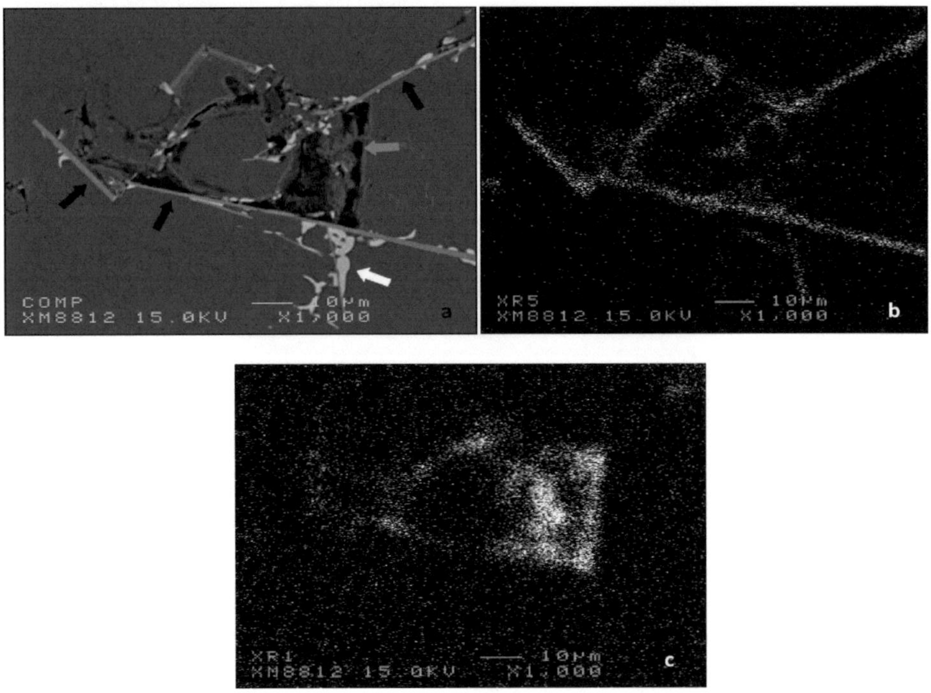

Figure 15.
(a) Backscattered electron image of oxide films mixed with β-Al_5FeSi platelets in D9 alloy, (b and c) X-ray maps of Fe and O_2, respectively in (a).

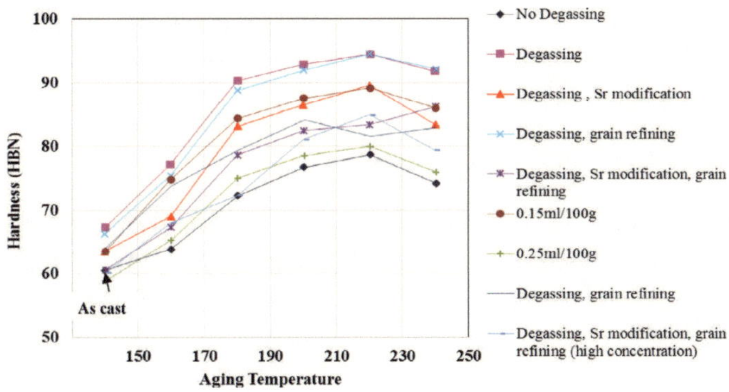

Figure 16.
Variation of the hardness of A319.0 alloy as a function of aging temperature and the applied melt treatment.

TiB_2 particles withing the gas pore as idendified by the X-ray electron maps presented in **Figure 11(b)** and **(c)**. As can be seen the X-ray map of Ti is well defined in the form of round spots whereas that of B is almost covering the entire area.

Figure 12 reveals another feature observed in D9 alloy the co-existence of TiB_2 (white arrows) alongwith SrO oxide films (or bifilms - bright spots) resulting in the formation of microporosity (black arrows). In earlier studies [30, 31], the authors' group showed the presence of microporosity in Sr-modified A319.0 alloy surrounded by SrO oxides as demonstrated in **Figure 13**.

The above-mentioned discussion contradicts the theory proposed by Argo and Gruzleski [32], who suggested that in unmodified alloys, the eutectic is characterized by its irregular solid/liquid interface, leading to entrapping of small pockets of liquid between advancing solidification fronts, causing the formation of microporosity, as shown in **Figure 14**. In Sr-modified alloys rather, a regular or planar interface results in widely dispersed larger porosity.

Figure 15(a) illustrates another source of porosity formation in D9 alloy. In this case, both the oxide films (gray arrows) and β-Al5FeSi platelets (black arrows) participated in the nucleation of irrgular pores [33–36]. Apparently the β-Al5FeSi platelets also act as a barrier blocking the propagation of the pore through the matrix. **Figure 15(b)** and **(c)** confirm the presence of the oxide films interacted with the β-Al5FeSi platelets. The white arrow in **Figure 15(a)** indicates the presence of a coarse Al_2Cu phase particle with no pores associated with it.

Figure 16 illustrates the variation in the hardness of the A319.0 alloy as a function of aging temperature and applied melt treatment. As can be seen, regardless of the applied treatment, all curves follow the same pattern since they are controlled by the precipitation of Al_2Cu phase particles. Apparently, the age hardening mechanism is completely independent of the melt treatment. In contrast, the hardness level revealed a clear response to the melt treatment used. When the alloy was not degassed, the hardness values were the lowest compared to other treatments, whereas the situation is inversed after proper degassing. Due to oxides associated with the grain refiner as well as the presence of SrO, the hardness dropped markedly. Increasing the hydrogen content to 0.25 ml/100 g (and hence a significant amount of porosity) led to hardness levels close to those obtained for mechanically stirred alloy (i.e., with no degassing).

4. Conclusions

Based on the results documented in the present work, the following conclusions may be drawn:

1. The use of the proper amount of grain refiner reduces the grain size of the as received alloy by about 80%. This refinement is mainly due to the large number of TiB_2 particles (50–100 nm).

2. Proper degassing can reduce the amount of porosity by about 90%.

3. Oxide films (or bifilms) either as a part of the added grain refiner, or due to oxidation of Sr, or both, would result in a significant increase in the percentage of porosity, thereby deteriorating the alloy strength, regardless of the applied aging treatment.

4. With respect to porosity formation, another parameter to consider is the alloy freezing range.

5. Although increasing the hydrogen level would lead to precipitation of round pores instead of the irregular shrinkage ones, such pores could markedly reduce the alloy strength to the level corresponding to the mechanically stirred alloy in the molten state.

6. Not all intermetallics contribute to the porosity formation except for the β-Al_5FeSi phase due to its precipitation in the form of intersecting platelets.

7. Age hardening behavior/pattern is governed by the hardening precipitates i.e., Al_2Cu and Mg_2Si, whereas the hardening level is a direct function of the applied melt treatment.

Author details

Dominique Gagnon[1], Agnes M. Samuel[1], Fawzy H. Samuel[1*], Mohamed H. Abdelaziz[2*] and Herbert W. Doty[3]

1 Département des Sciences Appliquées, Université du Québec à Chicoutimi, Chicoutimi, Canada

2 Département PEC, Universite' Française d'E' gypte, Ville Shorouk, Le Caire, Egypt

3 General Motors Materials Engineering, 823 Joslyn Ave, Pontiac, MI 48340, USA

*Address all correspondence to: fhsamuel@uqac.ca
and mohamed.abdelaziz@ufe.edu.eg

References

[1] Li XZ, Bian XF. Behavior of Hydrogen in Superheated Aluminum and its Alloys Melt. Materials Science Forum. 2000;**331-337**:209-214

[2] Anyalebechi PN. Hydrogen-induced gas porosity formation in Al–4.5 wt% Cu–1.4 wt% Mg alloy. Journal of Materials Science. 2013;**48**:5342-5353

[3] Jang Y, Jeong Y, Yoon C, Kim S. Fatigue life prediction for porosity-containing cast 319-T7 aluminum alloy. Metallurgical and Materials Transactions A. 2009;**40A**:1090

[4] Roy NR, Samuel AM, Samuel FH. Porosity formation in AI-9 Wt pct Si-3 Wt pct Cu alloy systems: Metallographic observations. Metallurgical and Materials Transactions A. 1996;**27**(2):415-429

[5] Roy NR, Zhang L, Louchez PR, Samuel FH. Porosity formation in Al-9 wt% Si3 wt% Cu-X alloy systems: Measurements of porosity. Journal of Materials Science. 1996;**31**(5):1243-1254

[6] A.M Samuel, F.H. Samuel and N.R. Roy, Porosity formation in Al-Si-Cu alloys: Role of intermetallics, in *Proceedings of the Light Metals Symposium, 35th Annual Conference of Metallurgists of CIM, Montreal, Quebec,* August *25-29, 1996*, p. 269.

[7] Bian XF, Zhang ZH, Liu XF. Effect of Strontium Modification on Hydrogen Content and Porosity Shape of Al-Si Alloys. Materials Science Forum. 2000;**331-337**:361-366

[8] Liu L, Samuel AM, Samuel FH, Doty HW, Valtierra S. Influence of oxides on porosity formation in Sr-treated Al-Si casting alloys. Journal of Materials Science. 2003;**38**:1255-1267

[9] Dahle AK, Nogita K, McDonald SD, Dinnis C, Lu L. Eutectic development and microstructure development in Al-Si alloys. Materials Science and Engineering A. December 2005;**413-414**(15):243-248

[10] F.H. Samuel, A.M. Samuel, H.W. Doty, S. Valtierra, On Porosity Formation in Al–Si–Cu Cast Alloys, Light Metals, 2017, pp. 243-255.

[11] C. M. Dinnis, M. O. Otte, A. K. Dahle, J. A. Taylor, The influence of strontium on porosity formation in Al-Si alloys, Metallurgical and Materials Transactions A, Vol. 35A (11), 2004, pp. 17-30.

[12] Samuel AM, Doty HW, Valtierra S, Samuel FH. Influence of Oxides on Porosity Formation in Sr-Treated Alloys. International Journal of Metalcasting. 2017;**11**:729-742

[13] Ye H. An overview of the development of Al-Si-Alloy based material for engine applications. Journal of Materials Engineering and Performance. 2003;**12**:288-297

[14] Campbell J, Tiryakioğlu M. Review of effect of P and Sr on modification and porosity development in Al–Si alloys. Materials Science and Technology. 2010;**26**(3):262-268. DOI: 10.1179/174328409X425227

[15] SkatTiedje N, Taylor JA, Easton MA. A new multi-zone model for porosity distribution in Al–Si alloy castings. Acta Materialia. 2013;**61**:3037-3049

[16] Miresmaeili SM, Shabestari SG, Boutorabi SMA. Effect of Melt Filtration on Porosity Formation in Sr-modified A356 Aluminum Alloy. International Journal of Cast Metals Research. 2003;**16**:541-548

[17] Miresmaeili SM, Shabestari SG, Boutorabi SMA. The Effect of Sr-modification Treatment on

Porosity Formation of Reduced Pressure 319 Al-alloy Castings. Canadian Metallurgical Quarterly. 2003;**42**:245-252

[18] Shabestari SG, Miresmaeili SM, Boutorabi SMA. Effects of Sr-modification and Melt Cleanliness on Melt Hydrogen Absorption of 319 Aluminum Alloy. Journal of Materials Science. 2003;**38**:1901-1907

[19] Taylor JA, Schaffer GB, StJohn DH. The role of iron in the formation of porosity in Al-Si-Cu-based casting alloys: Part I. Initial experimental observations. Metallurgical and Materials Transactions. 1999;**30**:1643-1650

[20] Khalifa W, Samuel AM, Samuel FH, Doty HW, Valtierra S. Metallographic observations of β-AlFeSi phase and its role in porosity formation in Al-7%Si alloys. International Journal of Cast Metals Research. 2006;**19**(3):156-166

[21] Samuel AM, Doty HW, Valtierra S, Samuel FH. Beta Al5FeSi phase platelets-porosity formation relationship in A319.2 type alloys. International Journal of Metalcasting. 2018;**12**(1):55-70

[22] Samuel AM, Doty HW, Valtierra S, Samuel FH. Influence of Oxides on Porosity Formation in Sr-Treated Alloys. International Journal of Metalcasting. 2017;**11**(4):729-742

[23] A.M. Samuel, F.H. Samuel, H.W. Doty and S. Valtierra, Porosity formation in Al-Si sand mold castings, International Journal of Metalcasting, Vol. 11(4), 2017, p pp. 812-822.

[24] Elsebaie O, Samuel AM, Doty HW, Valtierra S, Samuel FH. On the Impact Properties and Fractography of Al–11%Si Casting Alloy. International Journal of Metalcasting. 2018;**12**(1):36-54

[25] Qiu D, Taylor JA, Zhang M-X, Kelly PM. A mechanism for the poisoning effect of silicon on the grain refinement of Al–Si alloys. Acta Materialia. 2007;**55**:1447-1456

[26] Nafisi S, Ghomashchi R. Grain refining of conventional and semi-solid A356 Al–Si alloy. Journal of Materials Processing Technology. 2006;**174**:371-383

[27] J. Campbell, The Solidification of Metals, Iron and Steel Institute, Publication 110, London, 1967, p. 18.

[28] Liu L, Mohamed AMA, Samuel AM, Samuel FH, Doty HW, Valtierra S. Precipitation of β-Al5FeSi phase platelets in Al-Si based casting alloys. Metallurgical and Materials Transactions A. 2009;**40**(10):2457-2469

[29] Mohanty PS, Samuel FH, Gruzleski JE. Experimental study on pore nucleation by inclusions in aluminum castings. AFS Transactions. 1995;**103**:555-564

[30] Samuel AM, Doty HW, Valtierra S, Samuel FH. Beta Al5FeSi phase platelets-porosity formation relationship in A319.2 type alloys. International Journal of Metalcasting. 2018;**12**(1):55-70

[31] L. Liu, A.M. Samuel, F.H. Samuel, H.W. Doty and S. Valtierra, Influence of oxides on porosity formation in Sr-treated Al-Si casting alloys, *Journal of Materials Science,* Vol. 38(6), 2003, pp. 1255-1267.

[32] Argo D, Gruzleski JE. Porosity in Modified Aluminum Alloy Castings. AFS Transactions. 1988;**96**:65-74

[33] J, A. Taylor, Metal related castability effects in aluminum foundry alloys, Cast Metals, Vol. 8(4), 1995, pp. 225-252.

[34] J.A. Taylor, G.B Schaffer, J. D.H. StJohn, The role of iron in the formation

of porosity in Al-SiCu-based casting alloys (Parts 1-3), *Metallurgical and Materials Transactions A*, Vol. 30A(6), pp. 1643-1650 (part 1), pp. 1651-1655 (part 2), pp. 1657-1662 (part 3), 1999.

[35] Otte M, McDonald S, Taylor J, StJohn D, Schneider W. Controlling porosity-related casting rejects: understanding the role of iron in Al-Si alloys. AFS Transactions. 1999;**107**:471-478

[36] C.M. Dinnis, J.M. Taylor, A.K. Dahle, D.H. StJohn, The role of manganese and iron in porosity formation in aluminium-silicon foundry alloys, *Light Metals 2003, Proceedings 42nd Annual Conference of Metallurgists of CIM*, pp. 483-497,Vancouver, Canada, 2003.

Chapter 8

Shrinkage Porosity in Steel Sand Castings: Formation, Classification and Inspection

Nawaz Mahomed

Abstract

In this Chapter, shrinkage porosity defects in steel castings are analysed, particularly for low carbon, high alloyed steels, which have applications in critical engineering components. It begins with the mechanisms for porosity formation within the solidification contraction phase of the casting cycle, highlighting the importance of feeder design. This is followed by characterisation of the solidification phase of steel alloys, including the evolution of phases, which is important in distinguishing between microstructure and porosity in microscopy analysis. A more detailed discussion of interdendritic feeding and mechanisms for shrinkage micro-porosity is then provided. This leads to the well-established interdendritic flow model and commonly-used thermal criteria for shrinkage porosity prediction. The discussions are then consolidated through the classification of shrinkage porosity in terms of formation mechanisms and morphology, and its causes relating to composition, design and process conditions. Finally, engineering standards for classification and inspection of porosity types and severity levels in steel castings are discussed. Throughout, basic design and process improvement approaches for improving melt feeding during solidification contraction is given, with emphasis on providing practical solutions for prediction and evaluation of shrinkage porosity defects in castings.

Keywords: Shrinkage porosity, solidification, peritectic steels, digital radiography inspection

1. Introduction

Steel castings used in high-performance applications, such as cast valve components used in power plants and automotive engines, are subject to stringent quality standards to ensure operational integrity. Such steels may vary from high carbon low alloy steels to low carbon high alloyed steels, the latter designed to withstand highly-corrosive environments and have low susceptibility to brittle fracture. Metal castings of all material and process types are prone to porosity, defined as internal voids or non-metallic inclusions, and classified as internal defects (similar to other internal defects, such as the evolution of unwanted material phases or inhomogeneities in the distribution of alloying elements due to segregation phenomena). Such porosity defects can severely impact the performance of castings, through crack initiation and propagation under fatigue loading, stress corrosion cracking, and reduced material strength and (hence) fatigue life, among other factors [1–3]. It is therefore common that allowable limits on porosity severity levels are dictated by standards or client specifications. At the same time, it is therefore crucial that foundries develop a clear understanding of porosity defects in (steel) castings and introduce suitable monitoring and improvement protocols in order to remain competitive.

Porosity can generally be categorised as gas porosity (entrapped and segregated gases), inclusions (sand and slag), shrinkage porosity, and hot tears (also known as hot cracks) [4]. Furthermore, these different types of porosity may present both at the micro and macro length scales. These porosity types and their length scales, together with their morphologies, provide clues as to their likely causes, allowing the introduction of mitigation measures by foundries.

This Chapter focuses on shrinkage porosity in steel sand castings. Sandcasting, in particular, is a process that depends on gravitationally pressurised flow, which makes it difficult to feed the solidification contraction of the metal. This makes sandcast components, which is the case for most steel components, prone to shrinkage porosity. In some cases, geometric modifications to part geometry can eliminate shrinkage porosity. However, in other cases, it is unavoidable, but can be reduced through a combination of geometric and process modifications.

During the casting cycle, the steel undergoes three types of contractions that need to be considered, as illustrated in **Figure 1**.

- Firstly, the thermal contraction in the liquid state. In some practices, the liquid melt pouring temperature T_p is kept higher to avoid early formation of skin freezing, allowing the metal to solidify directionally inward from the surface. However, it is important to establish whether higher pouring temperatures lead to higher or lower shrinkage porosity. In any case, liquid melt contraction is not an issue in foundries, since its low value ($< 1\%$) is easily compensated through liquid feeding from the feeders. The liquid melt density as a function of temperature T $[K]$ and carbon content C [%wt] can be determined using the empirical relationship [5]:

$$\rho = (7.10 - 0.0732C) - (8.28 - 0.874C) \times 10^{-4}(T - 1823) \quad (1)$$

and the density of solid steel can be determined using Thermo-Calc or similar computational thermodynamic systems.

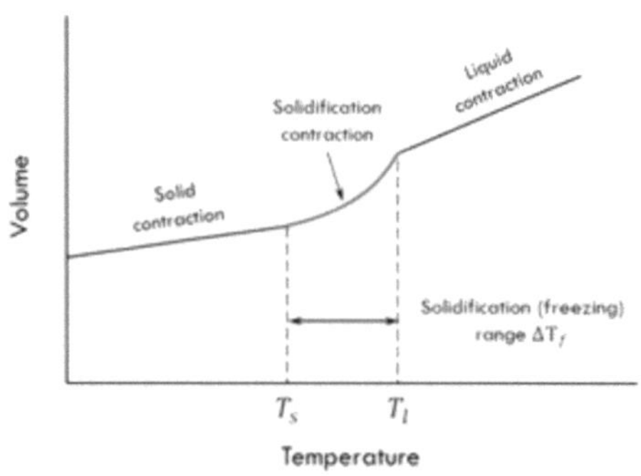

Steel properties: A216: WCB (Thermo-Calc)

Temperature (°C)		Density (kgm^{-3})	
Pouring T_p	1530	ρ_p	7094
Liquidus T_l	1495	ρ_l	7122
Solidus T_s	1430	ρ_s	7380
Ambient T_a	25	ρ_a	7820

Figure 1.
Contraction of steel during a casting cycle for a low carbon steel (A216 WCB with C = 0.3%, $T_{liquidus}$ = 1495°C, $T_{solidus}$ = 1430°C): liquid contraction = 0.4%; solidification contraction = 3.6%; solid contraction = 6.0%) (Data from Thermo-Calc and [5]).

- Secondly, the contraction during solidification in the mushy state, when the liquid metal transforms to the initial solid state microstructure. Contractions for some pure metals at their respective freezing temperatures are given in [4]. In the case of pure Fe, the solidification contraction was found to be 3.16% at the phase transformation temperature of 1536°C. However, as the carbon content increases, it increases to up to 5% for low carbon steels. This is related to the decrease in the solidus temperature as the carbon content increases (see phase diagram in **Figure 2**). The importance of solidification contraction is two-fold: (i) the need for an efficient feeder system to compensate for solidification contraction, and (ii) the formation of shrinkage porosity as a result of failure to feed the mushy zone – this is unavoidable due to coherency of the dendritic structure when the mushy zone reaches a particular solid fraction, leading to entrapped interdendritic liquid

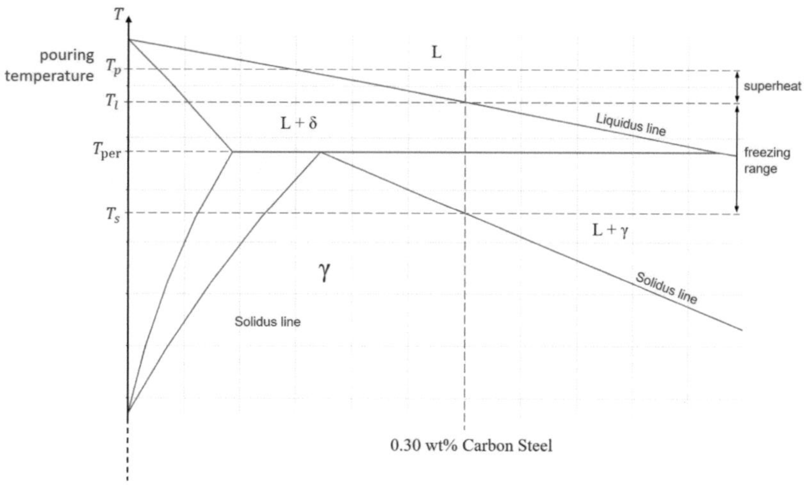

Figure 2.
Temperature profile during the solidification stage of a typical low carbon steel.

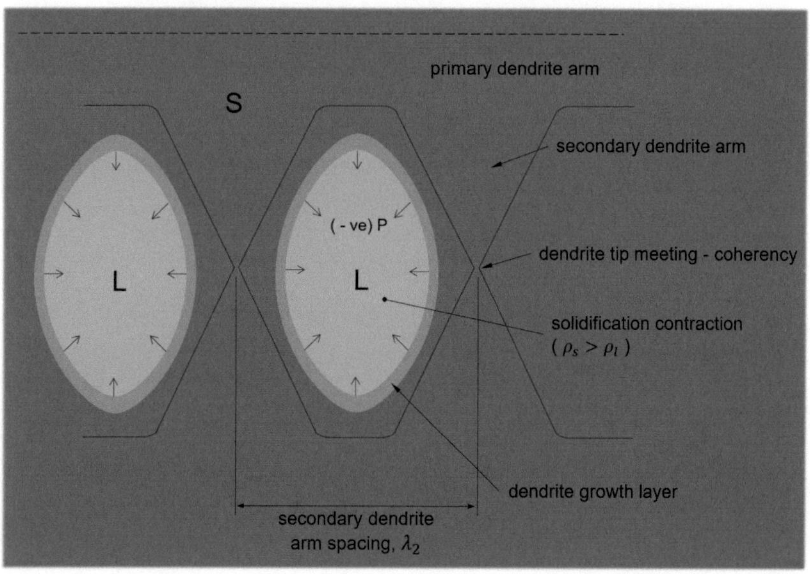

Figure 3.
Mushy zone coherency during solidification shrinkage.

melt, which results in either pore nucleation (due to volumetric shrinkage) or solid collapse (due to negative pressure), as shown in **Figure 3**. As will be shown later, the freezing range (difference between the liquidus and solidus temperatures) is an important factor in determining the morphology of shrinkage pores.

- Thirdly, the solid state contraction of the metal. Even though solid-state solidification values may be used in foundries (especially for pattern design), metals are rarely free to contract due to various microstructural constraints – this leads to internal stresses, and where these stresses overcome resistance, the material can undergo plastic or viscoplastic strain to adapt to the thermal and transformational density changes. Hence, the part may be larger than predicted. These constraints can lead to localised cold cracking (initiated from hot tears during solidification contraction).

2. Feeder System Design for Solidification Contraction

Attention is focused on the second stage, i.e. solidification contraction in the mushy zone, which is the root cause of shrinkage porosity. Initially, as stated above, a feeder system is required to compensate for solidification contraction, but also to ensure that shrinkage porosity formation is minimised. This requires that the feeder/s (a) must contain sufficient melt volume to compensate for the volume contraction of the part, and, more importantly, (b) solidify later than the casting (see **Figure 4**). The latter effectively implies that the thermal centre of the total casting should eventually migrate to the feeder/s, where high porosity formation due to hotspots or in the form of pipe shrinkage will be experienced.

Applying the two requirements for feeder design, firstly, the volume of the feeder/s is given by:

$$V_f \approx \alpha V_c$$

where α = shrinkage fraction, V_f = feeder/s volume and V_c = cavity volume. Taking account of shrinkage within the feeder itself:

$$V_f - \alpha V_f = \alpha V_c$$

However, feeders in sandcasting moulds have low levels of feeding efficiency due to gravitational feeding and low design heads, giving:

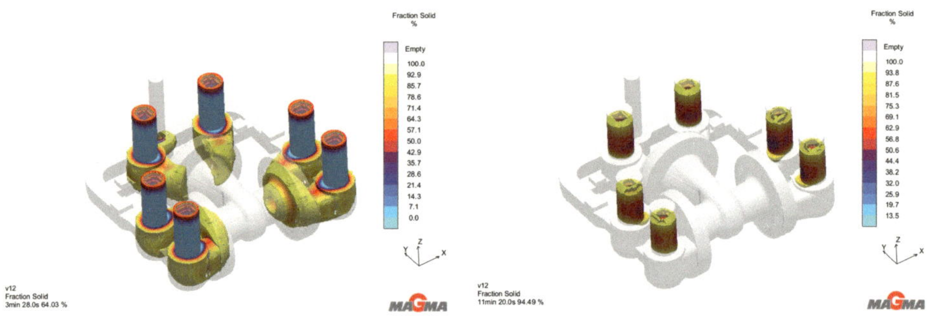

Figure 4.
Solid fraction predictions for a valve body casting at two different times, showing the feeders (with exothermic sleeves) solidifying later than the casting.

$$eV_f = \alpha(V_c + V_f) \tag{2}$$

where the feeder efficiency e is generally taken to be in the range of 15–30%. It is necessary to balance the feeder efficiency with feeder size to avoid high waste and energy consumption. Secondly, the solidification time of the feeder/s must be greater than that of the casting, typically by a "safety factor" of 1.2, giving:

$$t_s|_{casting} < t_s|_{feeder} \implies t_s|_{feeder} = 1.2t_s|_{casting} \tag{3}$$

The well-known Chvorinov's equation allows the above relation to be expressed in terms of the mould geometry, and can be expressed as:

$$t_s = B\left(\frac{V}{A}\right)^n \tag{4}$$

where V is the casting volume, A is the surface area of the casting through which heat is conducted, n is an empirical exponent (equal to 2 for simply shaped castings in silica sand moulds), and B is the moulding constant which is dependent on (a) process parameters (the equilibrium solidification temperature T_m, the initial temperature T_0 of the mould, and the superheat $\Delta T_s = T_{pour} - T_m$, all in °K), (b) mould material properties (thermal conductivity k in W·m^{-1}·K^{-1}, density ρ in kg·m^{-3}, and specific heat capacity c in J·kg^{-1}·K^{-1}), and (c) casting metal properties (latent heat of fusion L in J·kg^{-1}, density ρ_m in kg·m^{-3}, and specific heat capacity c_m in J·kg^{-1}·K^{-1}), given by:

$$B = \left[\frac{\rho_m L}{T_m - T_0}\right]^2 \left[\frac{\pi}{4k\rho c}\right]\left[1 + \left(\frac{c_m \Delta T_s}{L}\right)^2\right] \tag{5}$$

Although Eq. (4) may appear complex, it is relatively straightforward to evaluate given the casting process parameters deployed and the materials used for the sandmould and the casting, the latter being readily available in metal casting databases. It is therefore relatively easy to estimate the solidification time for a specific cast geometry. However, in the design of the feeder system, this is not necessary, since substituting Eq. (3) into Eq. (2) simply requires that:

$$\frac{V_f}{A_f} = 1.2\frac{V_c}{A_c} \implies m_f = 1.2m_c \tag{6}$$

where the modulus m has been introduced. Solving Eqs. (1) and (5) simultaneously for a given mould cavity allows the dimensioning of the system of feeders.

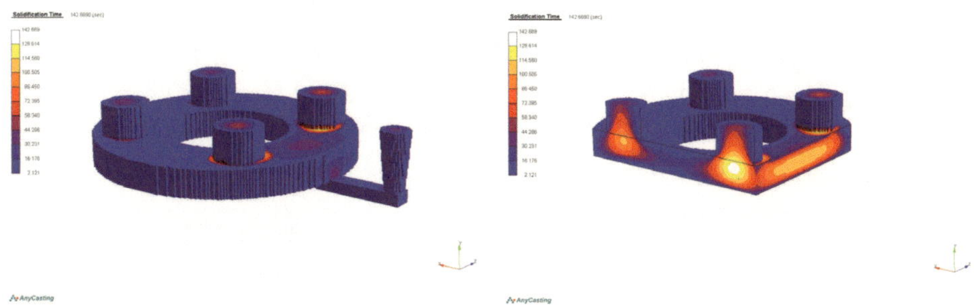

Figure 5.
Hotspot formation in the feeder junction regions and the gate area of a steel disc casting for a modulus ratio$m_f/m_c = 1.2$.

The effect of the modulus ratio m_f/m_c on shrinkage porosity is quite significant – with no feeding (i.e. modulus ratio = 0), shrinkage porosity as high as 7% can result. However, whilst modulus ratios higher than 1.2 can significantly reduce shrinkage porosity, it will not be entirely eliminated due to interdendritic phenomena. That is, the inclusion of oversized feeders does not guarantee sufficient pressure differentials to feed the solidifying dendritic network. This will be the subject of further investigation in the next section.

Furthermore, the modulus ratio of 1.2 does not guarantee the formation of hotspots in the junction region between the part and the feeder, as shown in **Figure 5**. This can be resolved through increasing the modulus ratio, albeit at the cost of production. However, a better approach would be to use exothermic sleeves around the feeders, as can be seen in **Figure 4**.

3. Interdendritic feeding

3.1 Solidification Characterisation of Steels

Before analysing interdendritic feeding, it is important to establish some important casting-related parameters of the steel during the solidification stage. These include the liquidus and solidus temperatures for the particular steel grade, so as to ascertain the superheat during pouring (i.e. the temperature difference between the pouring temperature (T_p) and the onset of solidification at the liquidus temperature (T_l)), and the freezing range (difference between the liquidus temperature (T_l) and solidus temperature (T_s)). In addition, the partition coefficient (k) and liquidus slope (m) are related to the evaluation of shrinkage porosity, as will be seen later.

Steels are generally alloyed, as well as post-cast heat-treated, to produce a desired microstructure (and hence specifically desired properties). Although the alloying levels (of chromium, nickel, etc.) may range from low (plain carbon steels) to high (stainless steels), the carbon composition plays a dominant role in determining dendrite size (and hence shrinkage porosity) of the solidifying melt [6, 7]. Of particular interest are low carbon steels, with carbon content in the range up to 0.3% wt. Such steels undergo a peritectic transformation during solidification at the peritectic temperature (T_{per}), at which δ-ferrite and liquid transforms into austenite. A number of empirical models can simply be used for determining the aforementioned temperatures for multi-component steels [8]. In this analysis, the binary (Fe-C) phase diagram of steel with C = 0.3%, as shown in **Table 1**, will be used to demonstrate these temperatures and their effect on shrinkage porosity[1].

The Fe-C phase diagram shown in **Figure 7** was generated using Thermo-Calc. For a steel with C = 0.3%wt, solidification starts at the liquidus temperature $T_l = 1515°C$ with the nucleation of BCC δ-ferrite, and proceeds until the peritectic temperature, $T_{per} = 1495°C$ is reached. This is referred to as the $L + \delta$ zone.

At this point, for the hyperperitectic (above C = 0.17%) steel, the phase fractions of value of δ-ferrite (of 0.10%C) can be calculated using the lever rule as follows:

$$\delta - ferrite : g_\delta = \frac{0.53 - 0.30}{0.53 - 0.10} = 0.53 \quad and \quad liquid : g_l^{per} = 1 - g_\delta^{per} = 0.47$$

---

[1] For an actual steel grade, the phase diagram will be slightly different due to the influence of the alloying elements. See **Figure 6** for the case of A216 WCB steel, showing an additional $L + \delta + \gamma$.

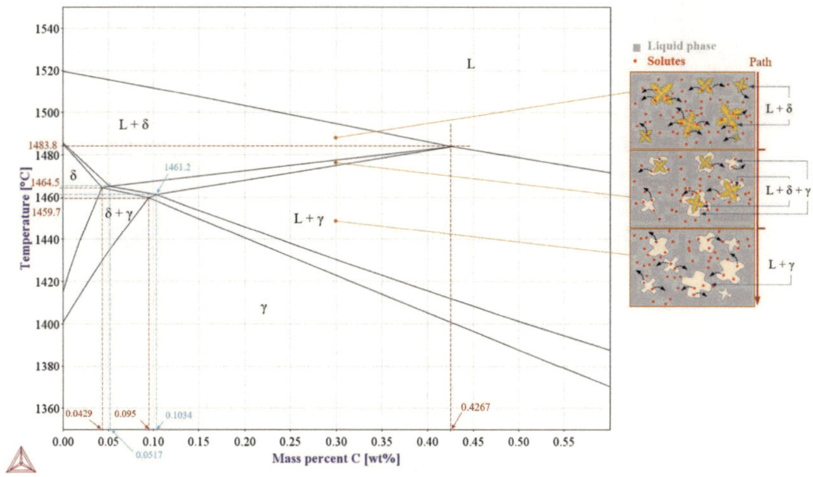

Figure 6.
Phase transformation of A216 WCB steel in the peritectic region (solidification zones), showing nucleation and growth of δ-ferrite and austenite at the respective transition zones across the freezing range from T_l = 1495°C to T_s = 1430°C (freezing range = 65°C), as well as solute rejection.

C	Mn	P	S	Si	Cu	Ni	Cr	Mo	V
As per ASTM A216 Material Standard									
0.30	1.00	0.035	0.035	0.6	0.30	0.50	0.50	0.20	0.03

Table 1.
Chemical composition (% wt) of A216 WCB cast steel.

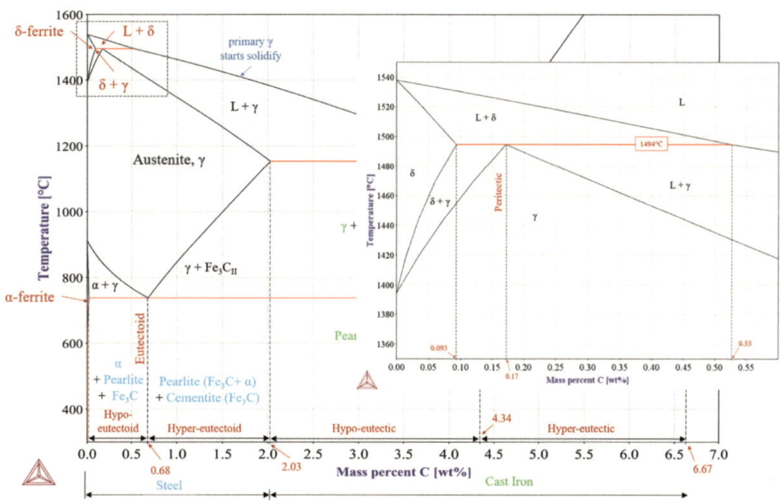

Figure 7.
Phase diagram of Fe-C cast steel, showing the peritectic region enlarged (generated using Thermo-Calc).

Now, for peritectic transformation to occur, using the lever rule at C = 0.17% (peritectic composition), the fractions of δ-ferrite (of 0.10% C) and liquid (of 0.53% C) are given as:

$$\delta - ferrite : g_{\delta}^{per} = \frac{0.53 - 0.17}{0.53 - 0.10} = 0.84 \quad and \quad liquid : g_{l}^{per} = 1 - g_{\delta}^{per} = 0.16$$

Hence, in our case of a hyperperitectic steel (C = 0.3%), there is a shortage of δ-ferrite to react with the amount of liquid fraction for a full peritectic transformation to austenite. The amount of liquid fraction that is transformed with the δ-ferrite fraction of 0.53 into austenite at the peritectic temperature is only $0.53/0.84 \times 0.16 = 0.10$. The resulting fractions of liquid and austenite just below the peritectic temperature are therefore given as:

$$liquid : g_l = 0.47 - 0.10 = 0.37 \quad and \quad austenite : g_\gamma = 1 - g_l = 0.63$$

The remaining liquid now transforms fully into austenite as the solid–liquid mixture cools down to the solidus temperature $T_s = 1470°C$. This is referred to as the $L + \gamma$ zone. (See [9] for an in-situ account of the solidification phase transformation of a peritectic steel). The freezing range ΔT_f is given as:

$$freezing\ range\ \Delta T_f = T_l - T_s = 1515°C - 1470°C = 45°C$$

It can be seen that hyperperitectic steels have longer freezing ranges compared to hypoperitectic steels, and high carbon steels (C > 0.53%) have even longer freezing ranges. In the latter, however, solidification starts with nucleation of FCC austenite instead of BCC δ-ferrite.

Further cooling of the solidified steel down to 800°C results in the transformation of austenite into α-ferrite (either allotriomorphic ferrite nucleating at the austenite grain boundaries or idiomorphic ferrite nucleating inside the austenite grains), until the eutectoid temperature of 740° C is reached. The remaining austenite in this hypoeutectoid steel then transforms into pearlite, a cooperative growth of α-ferrite and cementite (Fe_3C), hence showing up as a lamellae microstructure.

In the case of the 0.30% C steel, the volume fractions of α-ferrite and pearlite (α-ferrite + cementite) can similarly be determined using the lever rule:

$$\alpha - ferrite : g_\alpha = \frac{0.77 - 0.30}{0.77 - 0.02} = 0.63 \quad and \quad pearlite : g_{pearlite} = 1 - g_\alpha = 0.37$$

These volume fractions would be the case if no micro-porosity was present during the solidification stage, which is not exactly possible, even under highly efficient feeding of the mushy zone. Entrapped interdendritic liquid is bound to occur, even at high levels of solid fraction coherency. Porosity shows up as black (almost zero density) areas on a micrograph, which may sometimes be confused for the lower density phase (α-ferrite), as shown on the micrographs in **Figure 8**. Hence, it is important to establish the phase fractions as above, and use this data

Figure 8.
Optical Micrograph of an A216 WCB cast steel sample (left) and the contrasted image showing the porosity (right).

to distinguish between actual phases formed and porosity on a micrograph, since $g_\alpha + g_{pearlite} + g_{pores} = 1$. This effectively means that the density of the steel will be lower than its theoretical density, due to the formation of shrinkage pores.

The melt pouring temperature (T_p) is often chosen without a clear understanding of its effect on macro and microstructural defects. Knowing the liquidus temperature (T_l) allows a better choice of T_p. It is known that higher values of T_p increases the amount of porosity in castings [10], and increases hot tear susceptibility (HTS). The former is due to the increased possibility of hotspots, and the latter due to high thermally-induced stresses in the solidifying melt. Lower pouring temperatures are therefore preferred. **Figure 2** summarises the important temperature parameters during the solidification stage.

As a final word in this Section, the liquidus slope m and partition coefficient k can likewise be determined from the phase diagram of the steel alloy. An in-depth discussion on these parameters is beyond the scope of this Chapter, except to note that these parameters affect the spacing between secondary dendrite arms (as will be shown in Section 3.2), and hence the level of interdendritic shrinkage porosity. In peritectic steels, the liquidus slopes (temperature versus solute composition gradients) are given as m^δ and m^γ for the $L + \delta$ and$L + \gamma$ zones respectively. Similarly, the partition coefficients for the solute in δ-ferrite and austenite are given by:

$$k_C^\delta = \frac{C_{s,C}^*}{C_{l,C}^*} \quad and \quad k_C^\gamma = \frac{C_{s,C}^*}{C_{l,C}^*} \tag{7}$$

where $C_{s,C}^*$ refers to the solute composition of carbon (subscript "C") in the solid (subscript s) within the multicomponent steel alloy, and $C_{l,C}^*$ is similarly defined. The partition coefficient indicates the extent of solute rejection during the solidification process. Low values of k implies high levels of solute rejection. This can lead to high levels of microsegregation and, hence, inhomogeneity in microstructure – this is beyond the scope of this Chapter. As regards shrinkage porosity, the rejection of solute from the solidifying phase into the liquid melt contributes to the remelting of smaller dendrite arms and, subsequently, coarsening of larger dendrite arms. This is related to the difference in curvatures between the larger and smaller dendrite arms. A more detailed examination of dendritic coarsening can be found in [11]. The liquidus slopes and partition coefficients are important parameters for determining the evolution of solid fractiong_s as a function of temperature through the mushy zone, in the solution of shrinkage porosity formation, as will be shown in the next Section.

3.2 Mushy zone feeding

The nucleation and growth of the solid phase in the mushy zone involves highly complex phenomena, not all of which can be adequately dealt with in this Chapter. The aim here will be limited to discussing the mushy zone solidification in the context of shrinkage porosity formation. **Figure 9** depicts a typical equiaxed growth of a solid dendrite in a liquid melt, depicting the growth of secondary dendrite arms with spacing λ_2, and the velocity of the interdendritic feeding flow $\boldsymbol{v}_l$. The velocity of the solidifying front is shown as $\boldsymbol{v}_T$. The accompanying solidification contracts by a factor of:

$$\beta = (\rho_s - \rho_l)/\rho_l \tag{8}$$

(known as the solidification shrinkage factor), where ρ_s and ρ_l are the solid and liquid densities respectively. Hence, it can be shown that the (localised) feeding velocity becomes $\boldsymbol{v}_l = -\beta \boldsymbol{v}_T$.

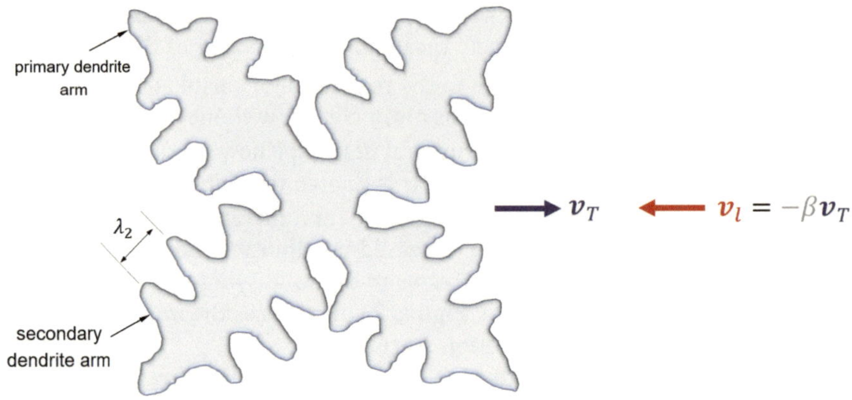

Figure 9.
Dendritic growth in a liquid melt.

The densities of liquid steel (at T_l) and solid steel (at T_s) across the mushy zone freezing range depends on alloying composition and phase transformations. The metal's density increases as the temperature reduces from T_l to T_s, with further volumetric shrinkage to room temperature. Additionally, phase transformations (such as that at the eutectoid temperature from FCC austenite to α-ferrite + cementite) change the degree of volumetric shrinkage. Refer to [12] for a more detailed discussion. The densities for different steel grades as a function of temperature can be determined using Thermo-Calc. In the case of A21 WCB steel, the value of $\beta = 0.033$ was established in **Figure 1**.

As the mushy zone solidifies, the liquid fraction (g_l) decreases at the expense of increasing liquid fraction (g_s), assuming that no pore fraction (g_p) forms ($g_l + g_s + g_p = 1$) as shown in **Figure 10**. This means that, at any time, the average localised feeding velocity across the mushy zone becomes $g_l \boldsymbol{v}_l$. This feeding velocity is dependent on the existence of a pressure gradient across the mushy zone, scaled by the permeability K of the mush zone and the dynamic viscosity of the liquid μ – related by the well-known Darcy equation (derived from conservation of momentum in a porous medium):

$$g_l \boldsymbol{v}_l = -\frac{K}{\mu}\left(\nabla \mathrm{p}_l - \rho_l \boldsymbol{g}\right) \tag{9}$$

where p_l is the liquid pressure and $\boldsymbol{g}$ is the gravitational acceleration.

The permeability of the porous mushy zone is modelled using the Kozeny-Carmen relation:

$$K\left(g_s(t_f), \lambda_2(t_f)\right) = \frac{\lambda_2^2(t_f)}{180} \frac{\left(1 - g_s(t_f)\right)^3}{g_s^2(t_f)} \tag{10}$$

where λ_2 is the secondary dendrite arm spacing (SDAS), dependent on the local solidification time t_f (the time for a point in the mushy zone to cool from the liquidus temperature T_l to the solidus temperature T_s). Eq. (10) indicates that coarser grain structures (larger λ_2) will increase the permeability K. Hence, coarser grain structure is preferred for higher permeability and improved feeding flow[2] [13]. In addition to

[2] It is noted that coarser grain structure can compromise the mechanical properties of a material, and it may be necessary to balance grain size for improved interdendritic flow and lower shrinkage porosity with mechanical properties, through controlling the solidification rate – see later.

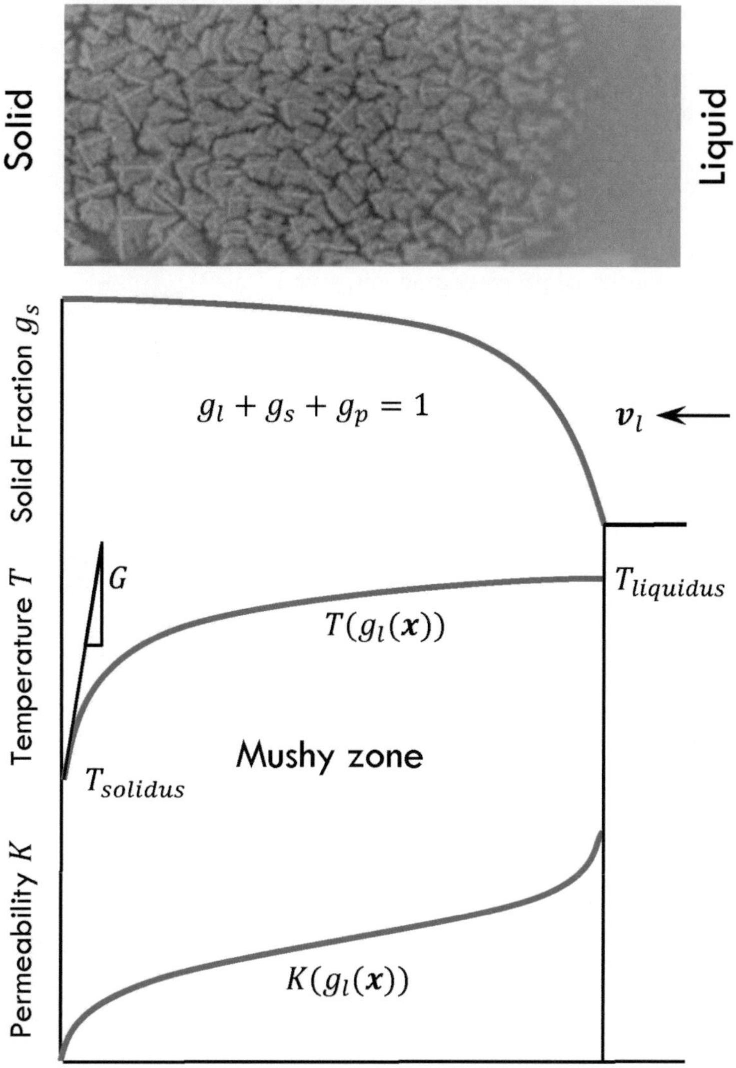

Figure 10.
Two-dimensional view of a typical equiaxed mushy zone: solid fraction, permeability and temperature variation.

local solidification time, this *coarsening* process is further compounded by what is referred to as the *remelting* of adjacent smaller dendrite arms, due to phenomena related to curvature differences and solute diffusion. A detailed analysis can be found in [11], except to mention that the SDAS can be determined anywhere in the mushy zone using a coarsening law:

$$\lambda_2(t_f) = K_{t_f}(t_f)^{\frac{1}{3}} \tag{11}$$

where K_{t_f} is a coarsening parameter derived experimentally. More complex models, based on solute diffusion, solidification phase transformation and the Gibbs–Thomson coefficient for the material, can be found [11]. Perhaps a more practical approach is to write Eq. (11) in terms of cooling rate $\dot{T}$ [14]:

$$\lambda_2(\dot{T}) = K_{\dot{T}}\dot{T}^{-\frac{1}{3}} \tag{12}$$

Ultimately, the interdendritic feeding velocity deep into the mushy zone drops rapidly (due to rapidly decreasing permeability), resulting in insufficient feeding flow and, hence, the onset of shrinkage porosity. Applying the conservation of mass $(\partial\rho/\partial t + \nabla\cdot(\rho\boldsymbol{v}) = 0)$ to the liquid domain:

$$\frac{\partial}{\partial t}(\rho_s g_s + \rho_l g_l) + \nabla\cdot(\rho_l g_l \boldsymbol{v}_l) = 0 \tag{13}$$

and substituting the Darcy Eq. (9), assuming that the gravitational pressure head in a sandcast is the actual pressure gradient, gives:

$$\underbrace{(\rho_s - \rho_l)\frac{\partial g_s}{\partial t} + \frac{\partial\rho_l}{\partial t}\left(1 - g_s - g_p\right) + g_s}_{\text{pore information}}\underbrace{\frac{\partial\rho_s}{\partial t} - \rho_l\frac{\partial g_p}{\partial t}}_{\text{solidification shrinkage}} = \underbrace{\nabla\cdot\left(\rho_l\frac{K}{\mu}(\nabla \mathrm{p}_l)\right)}_{\text{feeding}} \tag{14}$$

where $g_l + g_s + g_p = 1$ has been used to introduce the pore fraction in place of the liquid fraction. This form of the conservation equations gives a sense of the terms responsible for capturing (a) the shrinkage porosity compensated by feeding flow (positive) and (b) the pore formation due to a lack of feeding flow (negative). Eq. (14) can be reformulated by introducing the Kozeny-Carmen relation (Eq. (10)) and assuming that the liquid and solid densities variations $(\partial\rho_l/\partial t$ and $\partial\rho_s/\partial t)$ are minimal across the freezing temperature range:

$$(\rho_s - \rho_l)\frac{\partial g_s}{\partial t} - \rho_l\frac{\partial g_p}{\partial t} = \nabla\cdot\left(\rho_l\frac{1}{\mu}\frac{K_T^2\dot{T}^{-\frac{2}{3}}}{180}\frac{(1-g_s)^3}{g_s^2}(\nabla \mathrm{p}_l)\right)$$

and introducing β from Eq. (8), a concise mathematical form for shrinkage pore formation in the mushy zone can be established:

$$\beta\frac{\partial g_s}{\partial t} - \frac{\partial g_p}{\partial t} = \nabla\cdot\left(\frac{1}{\mu}\frac{K_T^2\dot{T}^{-\frac{2}{3}}}{180}\frac{(1-g_s)^3}{g_s^2}(\nabla \mathrm{p}_l)\right) \tag{15}$$

In practice, though, shrinkage porosity prediction is often based on a quantitative criterion, based on the scaling of the pressure gradient given in Eq. (15). By substituting for solid fraction g_s as a function of temperature (see **Figure 10**), using a microsegregation model (such as the lever rule or the Gulliver-Scheil model), it is possible to integrate Eq. (15) to obtain such a scaling parameter, and use this as a *criterion* for shrinkage pore formation. The scaling parameter, known as the Niyama criterion, given by:

$$N_y = \frac{G}{\sqrt{\dot{T}}} \tag{16}$$

where G is the temperature gradient at the solid-melt interface and $\dot{T}$ is the cooling rate, is currently the most widely used criterion for porosity prediction in metal casting. Niyama et al. [15] initially used this criterion to study porosity formation in steel castings, and concluded that (macro) porosity occurs when $N_y < 1$, i.e. low temperature gradient G and/or high cooling rate $\dot{T}$. Carlson and Beckermann [16] investigated the use of the Niyama criterion for shrinkage porosity in nickel alloy castings by simulating the filling and solidification and correlating the Niyama criterion with (micro- and macro-) porosity-containing areas in the

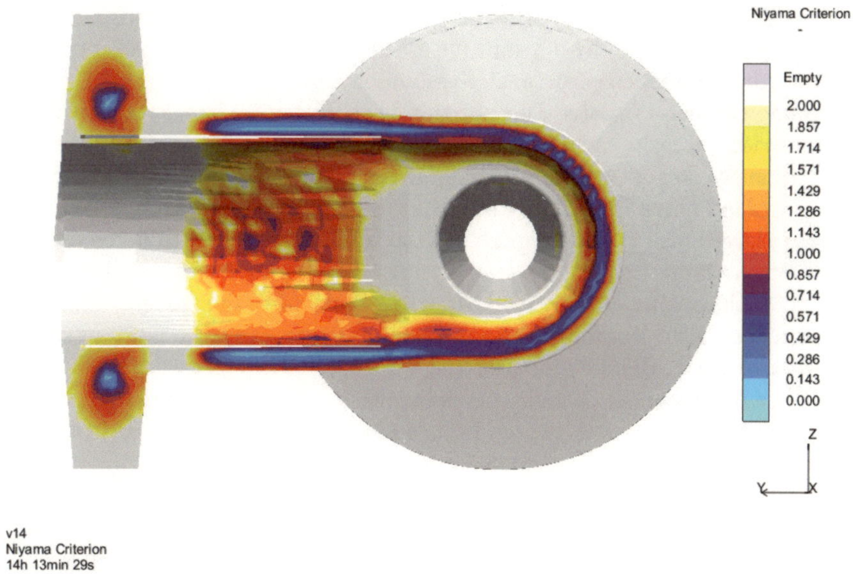

Figure 11.
Distribution of Niyama criterion values for a steel casting, showing macroporosity ($N_y < 1$) and microporosity ($N_y < 1$) in critical areas.

actual castings. They found that macroporosity (visible on radiographs) correlated to values of Niyama criterion $N_y < 1$, but also found that microporosity occurs at higher values of Niyama criterion $N_y < 2$ for nickel based alloys. Hence, critical areas in a steel casting should have values of Niyama criterion of at least $N_y > 2$ to be a sound casting (**Figure 11**).

Although the Niyama criterion, and other similar thermal criteria, such as that proposed by Lee et al [17] and Suri et al [18], are only quantitative indicators, it allows process control interventions to reduce shrinkage porosity formation. From the Niyama criterion (Eq. 16), it is clear that reducing the cooling rate ($\dot{T}$) will result in higher values of N_y and, hence, lower levels of shrinkage porosity - as mentioned before, this is due to coarser grain structures with higher permeability (i.e. improved feeding flow). Practically, this may involve preheating the mould, or/and using silica sand with lower thermal conductivity.

Furthermore, high thermal gradients (**G**), will also lead to higher values of N_y. From **Figure 10**, it is clear that lower freezing range allows (i.e. narrower mushy zones) will provide higher values of **G**, resulting in more columnar dendritic growth with higher permeability, as opposed to equiaxed growth with lower permeability.

4. Classification of Shrinkage Porosity Types

In the previous sections, the mechanisms for shrinkage porosity formation were discussed, from a thermal perspective – that is, during solidification contraction of the mushy zone. In this section, the various types of shrinkage porosity, based on morphology, will be classified. In doing so,, it is important to understand their likely causes, at least from a foundry perspective. A useful approach is to look at these causes in terms of design (both part and mould design), process and material factors.

At the outset, it is important to distinguish between micro and macro shrinkage porosity. A common misconception is to make this distinction purely based on length scale, for which there are different interpretations of what constitutes a micro or macro pore. A more substantive approach is to base this distinction on the microstructural and macrostructural phenomena involved.

Shrinkage Macro-Porosity. Part and mould geometry are largely responsible for shrinkage macro-porosity formation. Sudden increases in geometric volume can lead to the formation of hotspots, or entrapped liquid melt that solidifies inward (as was demonstrated in **Figure 5**). In such cases, the morphology can range from large spherical pores to large linear cracks (**Figure 12**), the latter also referred to as cold cracks.

Hotspots are large (macro) regions of entrapped liquid, surrounded by a frozen skin layer. Such regions initially lead to surface sinks until the frozen skin layer develops sufficient resistance to deformation, giving rise to large internal shrinkage pores. In cases where the hotspot is partially exposed to low-pressure liquid feeding, such as close to a gate, the porosity takes on either a layered morphology, referred to as layered shrinkage porosity (see **Figure 13**), or a pipe morphology, referred to as pipe shrinkage porosity.

Hot tears, as opposed to hotspot-induced shrinkage porosity, occur as a result of deformation caused by thermally-induced stresses. The volume fraction at which the mushy zone develops resistance to deformation, due to the coalescence of dendrite arms (see **Figure 10**), is referred to as the coherency point. The solid

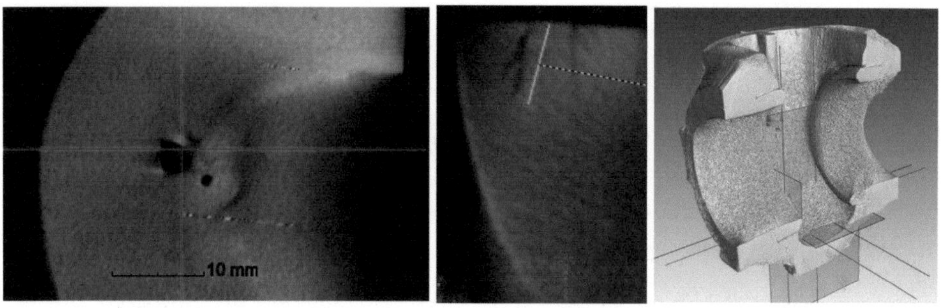

Figure 12.
Formation of macro shrinkage pores due to hotspot (left) and hot tearing / cold cracking (centre) located using computed tomography (CT) scanning (right).

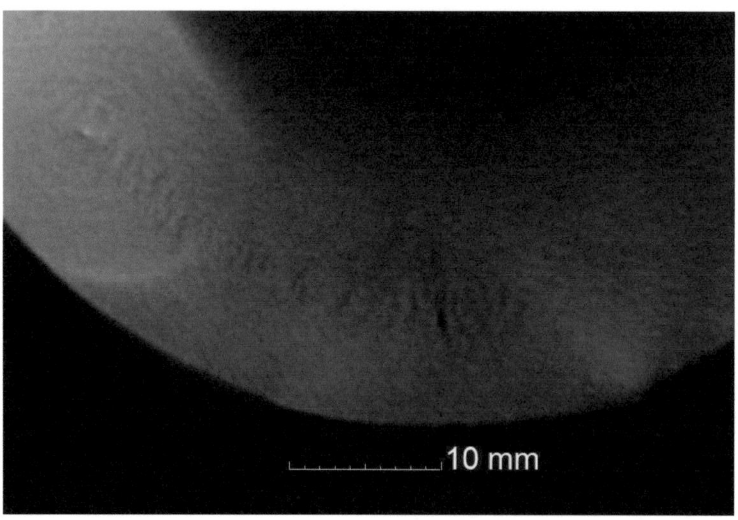

Figure 13.
Formation of layered macro shrinkage porosity due to a hotspot partially exposed liquid feeding.

fraction at which this occurs depends on factors such as solidification rate $\dot{T}$, the alloy freezing range and the diffusion of solute, and is typically in the region of $g_s = 0.8$ to $g_s = 0.95$ [11, 19]. At this point in the evolution of solid fraction, resisting stresses develop. Coupled with low permeability (i.e. lack of feeding), especially in low-freezing range alloys in which coarser dendritic structures evolve, solidification contraction of intergranular liquid can result in hot tearing. Although hot tearing is a microstructural phenomenon [20], it shows up as large intergranular cracks at the macrostructural scale.

Improving the pressurisation of the mould, through a combination of part geometry modifications and feeder design, can reduce or even eliminate the occurrence of hotspots, hot tears and layered porosity. In many cases, part design is often overlooked – this includes reducing sudden expansions in part volume, and introducing tapered flow regions in the part [21, 22] to improve feeding flow.

Shrinkage Micro-Porosity. As opposed to shrinkage macro-porosity, which is due to macrostructural effects such as hot spots and skin-freezing, micro-porosity is due to interdendritic shrinkage of entrapped liquid melt, nucleating mostly between secondary dendrite arms. As discussed in Section 3, the SDAS depends on the cooling rate during solidification, as well as the freezing range [14], the latter influenced by the carbon content of the steel [6], or, in general, chemical compositions of the major alloying elements. Depending on these factors, the SDAS can vary from as low as 50 μ m to as high as 700 μm [7], thus influencing the size of shrinkage micro-pores. For example, for a low carbon steel of 0.19%wt C, the size of the SDAS was found to be in the range of 67–311 μm, for cooling rates varying from 10 to 0.1°C/s [13].

Hence, the length scales of shrinkage micro-porosity may be as low as a few microns, increasing to multiple values of the SDAS in the case of interconnected interdendritic melt. Furthermore, three different pore morphologies can occur: linear, feathery and sponge, depending largely on the freezing range of the alloy and the cooling rate, as previously discussed. The different morphologies present different challenges with regards to part quality, with linear micro-porosity being classified as more severe as compared to sponge micro-porosity (as will be discussed further in the next section). Whilst the choice of alloy in many cases is out of the control of the foundry (and depends on client specifications based), understanding the mechanisms for shrinkage micro-porosity in terms of alloy freezing range is important in assessing expected part quality. Foundries can, though, control the cooling rate of the casting for a desired microstructure.

Importantly, the casting process parameters that prevail during the solidification phase have a direct impact on mechanical properties, such as strength and ductility (as a result of grain size), as well as microsegregation, although the latter could be remedied through solution heat treatment. For example, lower solidification rates lead to coarser microstructure (i.e. larger grain size), and hence improved interdendritic feeding. This can also improve material homogeneity (due to back-diffusion of solute); however, it will result in reduced ductility of the material (due to the coarser grain structure). Hence, the optimisation of casting and process parameters for reducing shrinkage porosity may need to be balanced with desired microstructural qualities, leading to multiple (and often conflicting) objectives during the optimisation process [10].

Shrinkage micro-porosity may present itself as three different distinct morphologies: Linear, Feathery and Sponge. The causes of these different morphologies are noted as follows:

- The morphology of shrinkage micro-porosity depends on the freezing range (a function of alloy composition), and solidification rate. The latter influences the

Figure 14.
Morphologies of linear (left) and feathery (right) shrinkage porosity.

permeability of the interdendritic region, in terms of SDAS, solid fraction distribution and point of coherency.

- In long-freezing-range alloys, due to inefficient liquid feeding of the interdendritic regions (as a result of pressure drops in the high-solid fraction region of the mushy zone), the resulting effect is that of a partially drained sponge, and hence the term "sponge porosity" for this defect (see **Figure 8**).

- In short freezing range alloys, the smoother solidification fronts lead to linear shrinkage defects, often of a feathery morphology. With low solidification rates, due to larger dendrite sizes and ensuing coarsening/remelting of neighbouring dendrite arms. A linear morphology prevails – **Figure 14**.

5. Porosity inspection

Procedures for porosity inspection in castings are specified under the relevant industry standard, depending on the material, thickness of the casting and the radiographic technique used. These procedures, based on the ASTM standards, are listed in **Table 2** for different part thicknesses.

In this Chapter, reference will be limited to the ASTM E2868 standard (Standard Digital Reference Radiographs for Steel Castings Up to 2 inches (50.8 mm) in Thickness). This is the equivalent of ASTM E446 for film radiography, both of which are similar to their heavy-walled counterparts. The approach developed here can therefore be applied to any of these standards. The ASTM E2868 standard provides a set of five digital reference radiographic images (nominal 250 kV X-rays) for each type of shrinkage porosity (CA - linear, CB – feathery, CC – sponge, CD – combined) from the lowest severity level 1 to the highest severity level 5 (as shown in **Figure 15** for the case of sponge shrinkage porosity), for comparison with production radiographs. The reference radiographs can be generated for a comparable resolution (pixel size) to that of the production radiographs.

	Thin-walled →	← Heavy-walled →	← Heavy-walled →
Standard (film) reference radiographs	ASTM E446	ASTM E186	ASTM E280
Standard digital reference radiographs	ASTM E2868	ASTM E3030	

Table 2.
ASTM standards for film and digital radiography of steel castings.

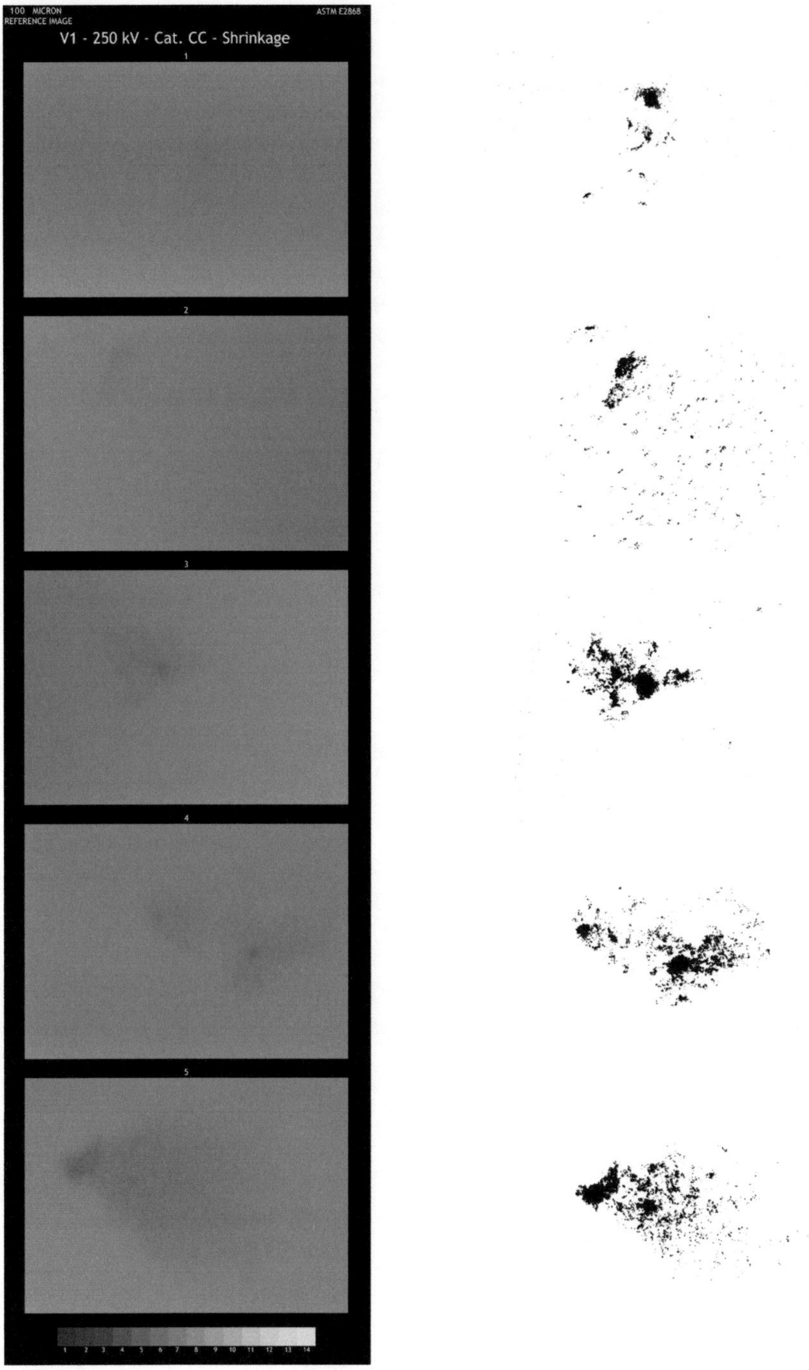

Figure 15.
Original (left) and contrasted (right) 100 μm resolution ASTM E2868 digital reference radiographs for shrinkage porosity Category CC (sponge), severity levels 1–5.

Production radiographs are commonly obtained using conventional X-ray techniques. Lately, X-ray Computed Tomography (CT) is gaining popularity for internal defect identification [23]. Reasons include the ability to determine pore morphologies in 3D, and resolutions as low as 1 μm (voxel size) suitable for micro-porosity detection. For comparison with reference radiographs as per the industry standards, 3D CT scans can be collapsed into equivalent 2D images, using a "thick slab" feature of CT scanning image analysis software [24].

The standard digital reference radiographs provide resolutions as low as 20 μm (pixel or voxel size) – however, it may not be possible to generate such high resolution production radiographs in practice due to equipment limitations, especially in the case of high density materials such as steels. Hence, a 100 μm resolution is seen as adequate for detection within the typical SDAS length scale (for ferrous metals).

Allowable porosity levels are often based on client specifications, but may also be part of quality standards associated with particular components. Such acceptance criteria, as in the case of high-performance valves for example, are shown in **Figure 16**.

The inspection process requires the area of the production radiograph to be prorated to an area of interest of the reference radiograph, as illustrated in **Figure 17**, and visually compared to establish the matching severity level.

Shrinkage Porosity Type	Category	Acceptable comparative radiograph (ASTM E446 / ASTM E2868)
Shrinkage - linear	CA	CA2
Shrinkage – feathery	CB	CB3
Shrinkage – sponge	CC	CC3
Shrinkage – combined	CD	CD3

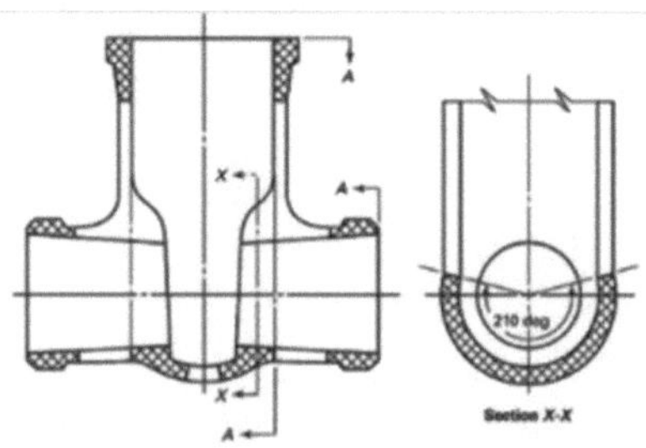

Figure 16.
Acceptance criteria for shrinkage porosity in steel valve castings of wall thickness ≤ 50.8 mm (ASME B16.34), showing the critical regions on the right.

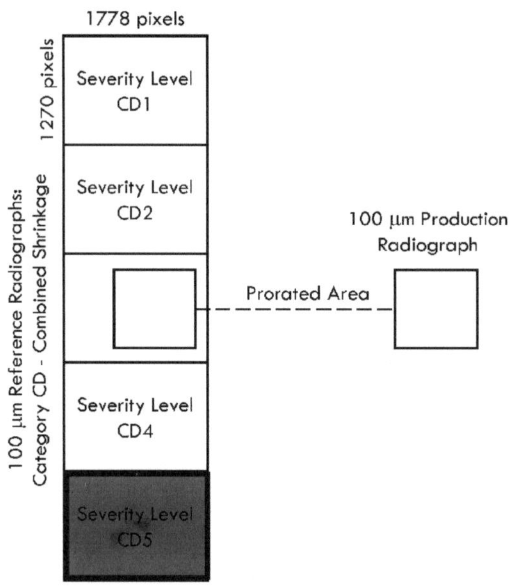

Figure 17.
Evaluating the porosity severity level by comparing the production radiograph with the reference radiographs of comparable resolution (e.g. 100 μm = 10 pixels/mm) from ASTM E2868, for a particular category of shrinkage porosity (e.g. CD - combined shrinkage porosity).

This inspection process is considered to be subjective, and other objective methods based on image processing have been suggested [25, 26], a subject for further development in the future.

6. Summary

An overview of shrinkage porosity is given, with reference to steel sand castings, from investigating the causes of shrinkage porosity to possible mitigation approaches, and finally the importance of understanding the technologies associated with the inspection and evaluation of shrinkage porosity severity levels. Further analysis of certain aspects of shrinkage porosity may be undertaken where an in-depth understanding is required. In summary, the following points are noted:

1. Shrinkage micro-porosity arises during the solidification contraction of an alloy. Hence, the steel's phase transition and density properties in the mushy zone are important in understanding the likely mechanisms and expected morphologies of shrinkage porosity.

2. Linked to this, the lack of interdendritic feeding of the mushy zone as its solid fraction increases beyond a certain point, referred to as the coherency point, leads to interdendritic pore nucleation. In large freezing range alloys, the result is a sponge morphology, migrating to feathery linear morphologies as the freezing range decreases (with decreasing carbon content).

3. However, the cooling rate is a determining factor for secondary dendrite arm spacing, and hence level of shrinkage porosity in an alloy. This gives the foundry a degree of control in avoiding the more severe linear porosity.

4. At the intergranular level, entrapped liquid melt can lead to hot tearing, due to the weak resisting forces of the coherent dendritic structure, succumbing to inward pressure. Hence, hot tearing may show up as macro-porosity in castings.

5. Shrinkage macro-porosity, on the other hand, is due to large entrapped liquid volume that cannot be adequately fed by the feeder system because of skin freezing. Whilst hotspots can easily be overcome through part and mould design modifications, narrow long sections may experience layered porosity due to declining pressurization of the mould as the skin layer thickens. Special attention should therefore be given to the design of the feeder system in such castings. This also goes for hotspots in the junction region between the casting and a feeder, which can lead to pipe shrinkage porosity extending into the part.

6. It was shown that certain thermal criteria may be used in practice in determining or reducing the level of shrinkage porosity, both micro-porosity and macro-porosity.

7. Finally, assessing the severity level of porosity in a casting is necessary in determining the acceptability of cast part quality. A range of standards have been issued, which provide the foundry with a benchmark in terms of acceptance criteria for the various types of shrinkage porosity.

Acknowledgements

The author wishes to thank Llewellyn Cupido for producing some of the figures and for generating the Thermo-Calc data used in the Chapter.

Author details

Nawaz Mahomed
Stellenbosch University, Stellenbosch, South Africa

*Address all correspondence to: nawaz@sun.ac.za

References

[1] Hardin RA, Beckermann C. Effect of Porosity on the Stiffness of Cast Steel. Metall Mater Trans A. 2007;**8**(12): 2992-3006. DOI: 10.1007/s11661-007-9390-4

[2] Ol'khovik E. Study of the Effect of Shrinkage Porosity on Strength of Low Carbon Cast Steel. IOP Conf Ser-Mat Sci. 2015;91(1). ISSN 1757–8981

[3] Sigl K., Hardin R.A., Stephens R., Beckermann C. Fatigue of 8630 cast steel in the presence of porosity. Int J Cast Metal Res. 2004;17(3):130–146. DOI: 10.1179/136404604225020588

[4] Campbell J. Castings. 2nd ed. Oxford: Butterworth-Heineman; 2003. 335 p. ISBN: 0750647906

[5] Jimbo, I., Cramb, A.W. The density of liquid iron-carbon alloys. Metall Mater Trans B. 1993;24B: 5–10. DOI: 10.1007/BF02657866

[6] Pierer R., Barnard C. On the influence of carbon on secondary dendrite arm spacing in steel. J. Mater. Sci. 2008;43:6938–6943. DOI 10.1007/s10853-008-2985-3

[7] El-Bealy M., Thomas B.G. Prediction of Dendrite Arm Spacing for Low Alloy Steel Casting Processes. Metall Mater Trans B. 1996;27B: 689–692. 10.1007/BF02915668

[8] Martiník O. *et al*. Eperimental and Theoretical Assessment of Liquidus, Peritectic Transformation, and Solidus Temperatures of Laboratory and Commercial Steel Grades. Journal of Phase Equilibria and Diffusion. 2019;40(1):93–103. DOI: 10.1007/s11669-019-00707-1

[9] Chen H. *et al*. Phase Transition of Peritectic Steel Q345 and Its Effect on the Equilibrium Partition Coefficients of Solutes. Metals. 2017;7(8):Article 288. DOI: 10.3390/met7080288

[10] Mahomed N., Kleynhans H.A. Reducing Shrinkage Porosity in High-Performance Steel Castings: Case of an ASME B16.34 Gate Valve Body: Part 2 — Simulation and Experimental Verification. Int J Metalcast. 2019;13(2): 463–472. DOI: 10.1007/s40962-018-0275-7

[11] Dantzig J.A., Rappaz M. Solidification. 1st ed. Lausanne: EPFL Press; 2009. 621 p. ISBN:978–2–940222-17-9

[12] Tesfaye Firdu F., Taskinen P. Densities of Molten and Solid Alloys of (Fe, Cu, Ni, Co) - S at Elevated Temperatures - Literature Review and Analysis. Aalto University Publications in Materials Science and Engineering, Espoo; 2010.

[13] Carlson K.D., Beckermann C. Prediction of Shrinkage Pore Volume Fraction Using Dimensionless Niyama Criterion. Metall Mater Trans A. 2009; 40:163–175. DOI: 10.1007/s11661-008-9715-y

[14] Zhao J., Yan P. The Effect of Cooling Rate of Solidification on Microstructure and Alloy Element Segregation of As Cast Alloy 718. In: Proceedings of the International Symposium on Superalloys and Various Derivatives; 17 June 2001; Pittsburgh. P. 133–140

[15] Niyama E., Uchida T., Morikawa M., Saito S. A Method of Shrinkage Prediction and its Application to Steel Casting Practice. Int. Cast. Met. J. 1982;7 (3): 52–63.

[16] Carlson K.D., Beckermann C. Use of the Niyama Criterion to Predict Shrinkage-related Leaks in High-Nickel Steel and Nickel-Based Alloy Castings. In: Proceedings of 62nd SFSA Technical and Operating Conference. 2008; Chicago. Steel Founders Society of America. Paper No. 5.6

[17] Lee Y. W., Chang E., Chieu C. F. Modeling of feeding behavior of solidifying Al-7Si-0.3Mg alloy plate casting. Metall. Mater. Trans. B. 1990;21 (4):715–722. DOI: 10.1007/BF02654250

[18] Suri V.K., Paul A.J., El-Kaddah N., Berry J.T. Determination of correlation factors for prediction of shrinkage in castings, Part I: Prediction of microporosity in castings: a generalised approach. AFS Trans. 138, 861–867, 1994

[19] Khalajzadeh V., Beckermann C. Simulation of Shrinkage Porosity Formation During Alloy Solidification. Metall Mater Trans A. 2020;51A:2239–2254. DOI: 10.1007/s11661-020-05699-z

[20] Ridolfi M.R. Hot Tearing Modeling: A Microstructural Approach Applied to Steel Solidification. Metall Mater Trans B. 2014;45B:1425–1438. DOI: 10.1007/s11663-014-0068-1

[21] Riedler M., Michelic S., Bernhard C. Formation of shrinkage porosity during solidification of steel: Numerical simulation and experimental validation. IOP Conf Ser-Mat Sci. 2016;143(1). DOI:10.1088/1757-899X/143/1/012035

[22] Monroe C.A., Gorsky D.A., Huff R. K., Grandhi R.V. Improving the Directional Solidification of Complex Geometries through Taper Addition. Int. J. Metalcast. 2014;8(3):23–27. DOI: 10.1007/BF03355587

[23] Yang Z., Kang J., Wilkinson D.S. Characterization of Pore Defects and Fatigue Cracks in Die Cast AM60 Using 3D X-ray Computed Tomography. Metallurgical and Materials Transactions B. 2015;46B:1577–1585. DOI: 10.1007/s11663-015-0370-6

[24] Mahomed N., Kleynhans H.A. Reducing Shrinkage Porosity in High-Performance Steel Castings: Case of an ASME B16.34 Gate Valve Body: Part 1 — Analysis, Techniques and Experimental Approach. Int J Metalcast. 2018;12(4): 919–926. DOI: 10.1007/s40962-018-0226-3

[25] Carlson K., Ou S., Hardin R., Beckermann C. Analysis of ASTM X-ray shrinkage rating for steel castings. Int. J. Cast Metals Res. 2001;14:169–183. DOI: 10.1080/13640461.2001.11819436

[26] Blair M., Monroe R., Hardin R., Beckerman C., A New Standard for Radiographic Acceptance Criteria for Steel Castings. In: Proceedings of the 62nd SFSA Technical and Operating Conference; 2008; Chicago. Steel Founders' Society of America. Paper No. 5.5